T0207545

Communications in Computer and Information Science 1527

More information about this series at https://link.springer.com/bookseries/7899

Gregor Rozinaj · Radoslav Vargic (Eds.)

Systems, Signals and Image Processing

28th International Conference, IWSSIP 2021
Bratislava, Slovakia, June 2–4, 2021
Revised Selected Papers

Editors
Gregor Rozinaj
Slovak University of Technology in Bratislava
Bratislava, Slovakia

Radoslav Vargic
Slovak University of Technology in Bratislava
Bratislava, Slovakia

ISSN 1865-0929 ISSN 1865-0937 (electronic)
Communications in Computer and Information Science
ISBN 978-3-030-96877-9 ISBN 978-3-030-96878-6 (eBook)
https://doi.org/10.1007/978-3-030-96878-6

This Springer imprint is published by the registered company Springer Nature Switzerland AG
The registered company address is: Gewerbestrasse 11, 6330 Cham, Switzerland

Preface

Signal processing is a phenomenon which is a principal tool for all applications based on computer technologies for an interaction with the real world. Data/signal retrieval, processing, and visualization comprise the general methodology in most scientific areas. From filter design, Fourier and other transforms, feature extraction, etc. through machine learning and system adaptation to user-oriented products like 5G networks, IoT, virtual teleportation, or tele-surgery operations, this is a very brief resume of the power of signal processing. Advanced signal processing is therefore a very complex topic with deeply structuralized content.

The International Conference for Systems, Signal and Image Processing (IWSSIP) is a well-established event with traditional participation of people from all continents. In 2021, the 28th IWSSIP conference was organized by the Slovak University of Technology, Bratislava, Slovakia, following the previous successful IWSSIP 2020 event in Rio de Janeiro, Brazil, and past conferences in various countries.

We are proud of having outstanding invited speakers who significantly increased the scientific quality of this event. Touradj Ebrahimi from École Polytechnique Fédérale de Lausanne, Switzerland, presented the state of the art in image compression standards based on artificial intelligence. Gabriel Miro-Muntean from Dublin City University, Ireland, gave a deep introduction to "Delivering High Quality Rich Media Content in Current Network Environment: Challenges and Solutions". Abir Hussain from Liverpool John Moores University, UK spoke about the "Detection and Localization of Objects Within Images Using Computer Vision and Machine Learning". We are deeply thankful to all three speakers for their valuable time and invited lectures.

Due to special COVID-19 pandemic regulations the presentations of all three keynote speakers, as well as the whole conference, was organized in a strictly online format with the stress on full interaction among IWSSIP participants. The whole conference was powered by underline.io and we would like to express our gratitude for their professional help with organizing the online event.

Despite the fact that one of the main goals of scientific conferences is to gather researchers working in similar areas in one place, there was still huge interest in the online IWSSIP 2021 event. We received 76 paper proposals from authors in 17 countries. The best 20 papers were accepted and selected for this publication. These papers are closely related to advances in signal processing. The orientation of presented papers shows the huge diversity and complexity of advanced signal processing.

We would like to thank all participants of IWSSIP 2021, as well as everyone who helped to make this conference successful.

June 2021

Gregor Rozinaj
Radoslav Vargic

Organization

Honorary Chairs

Branka Zovko-Cihlar	University of Zagreb, Croatia
Pavol Podhradský	Slovak University of Technology, Slovakia

General Chair

Gregor Rozinaj	Slovak University of Technology, Slovakia

Program Committee Chair

Radoslav Vargic	Slovak University of Technology, Slovakia

Steering Committee

Aura Conci	Universidade Federal Fluminense, Brazil
Mislav Grgić	University of Zagreb, Croatia
Sonja Grgić	University of Zagreb, Croatia
Fabiana Leta	Universidade Federal Fluminense, Brazil
Panos Liatsis	Khalifa University of Science and Technology, United Arab Emirates
Pavol Podhradsky	Slovak University of Technology, Slovakia
Snjezana Rimac-Drlje	University of Osijek, Croatia
Gregor Rozinaj	Slovak University of Technology, Slovakia
Markus Rupp	Technische Universität Wien, Austria
Radoslav Vargic	Slovak University of Technology, Slovakia
Branka Zovko-Cihlar	University of Zagreb, Croatia

Program Committee

Narcis Behlilovic	University of Sarajevo, Bosnia and Hercegovina
Ángel Sánchez Calle	Universidad Rey Juan Carlos de Madrid, Spain
Aura Conci	Universidade Federal Fluminense, Brazil
Jan Cornelis	Vrije Universiteit Brussel, Belgium
Žarko Čučej	University of Maribor, Slovenia
Marek Domański	Poznań University of Technology, Poland
Touradj Ebrahimi	EPFL, Switzerland

Irena Galic	University of Osijek, Croatia
Dušan Gleich	University of Maribor, Slovenia
Mislav Grgić	University of Zagreb, Croatia
Sonja Grgić	University of Zagreb, Croatia
Yo-Sung Ho	Gwangju Institute of Science and Technology, South Korea
Ebroul Izquierdo	Queen Mary University of London, UK
Dimitrios Karras	National and Kapodistrian University of Athens, Greece
Erich Leitgeb	Graz University of Technology, Austria
Fabiana Leta	Universidade Federal Fluminense, Brazil
Panos Liatsis	Khalifa University of Science and Technology, United Arab Emirates
Rastislav Lukac	Intel Corporation, Canada
Galia Marinova	Technical University of Sofia, Bulgaria
Marta Mrak	Queen Mary University of London, UK
Peter Planinšič	University of Maribor, Slovenia
Pavol Podhradsky	Slovak University of Technology, Slovakia
Snježana Rimac-Drlje	University of Osijek, Croatia
Gregor Rozinaj	Slovak University of Technology, Slovakia
Markus Rupp	Technische Universität Wien, Austria
Ryszard Stasiński	Poznan University of Technology, Poland
Boris Šimak	Czech Technical University in Prague, Czech Republic
Rodica Tuduce	University Politehnica of Bucharest, Romania
Ján Turán	Technical University of Košice, Slovakia
Radoslav Vargic	Slovak University of Technology, Slovakia
Stamatis Voliotis	Technological Educational Institute of Chalkida, Greece
Krzysztof Wajda	AGH University of Science and Technology, Poland
Drago Zagar	University of Osijek, Croatia
Theodore Zahariadis	National and Kapodistrian University of Athens, Greece
Branka Zovko-Cihlar	University of Zagreb, Croatia

Additional Reviewers

Abreu, Raphael
Aguilera, Cristhian A.
Araújo, José Denes
Aung, Zeyar
Bergamasco, Leila
Bergo, Felipe
Bezerra, Eduardo
Bozek, Jelena
Bravenec, Tomas
Bujok, Petr

Burget, Radim
Casaca, Wallace
Ciarelli, Patrick Marques
Copetti, Alessandro
Costa, Tales Fernandes
Čepko, Jozef
Davídková Antošová, Marcela
Devamane, Shridhar
Gonçalves, Vagner Mendonça
Habijan, Marija
Haddad, Diego Barreto
Henriques, Felipe
Hocenski, Zeljko
Hrad, Jaromír
Jakóbczak, Dariusz Jacek
Juhár, Jozef
Kačur, Juraj
Karwowski, Damian
Kominkova Oplatkova, Zuzana
Körting, Thales Sehn
Kos, Marko
Kultan, Matej
Laguna, Juana Martinez
Latkoski, Pero
Lima, Alan
Londák, Juraj
Lopes, Bruno
Lopes, Guilherme Wachs
Lourenço, Vítor
Malajner, Marko
Mandic, Lidija
Marana, Aparecido Nilceu
Marchevský, Stanislav
Markovska, Marija
Matos, Caio
Medvecký, Martin
Minárik, Ivan
Mocanu, Stefan
Mustra, Mario
Nyarko, Emmanuel Karlo
Paiva, Anselmo
Papa, Joao Paulo
Polak, Ladislav
Prinosil, Jiri
Rakús, Martin
Rodriguez, Denis Delisle
Rybárová, Renata
Silva, Aristófanes
Silvestre, Santiago
Slanina, Martin
Sousa De Almeida, Joao Dallyson
Sousa, Azael Melo E.
Tcheou, Michel
Toledo, Yanexis Pupo
Trúchly, Peter
Veras, Rodrigo
Vitas, Dijana
Vlaj, Damjan
Vukovic, Josip
Zamuda, Ales
Zeman, Tomas

Organizing Committee

Zuzana Bruncliková, Slovakia
Lucia Hlinková, Slovakia
Juraj Kačur, Slovakia
Juraj Londák, Slovakia
Ivan Minárik, Slovakia
Pavol Podhradský, Slovakia
Šimon Tibenský, Slovakia
Marek Vančo, Slovakia
Radoslav Vargic, Slovakia

Contents

Advanced Signal Processing

Segmentation and Quantification of Bi-Ventricles and Myocardium Using 3D SERes-U-Net

Marija Habijan[1(✉)], Irena Galić[1], Hrvoje Leventić[1], Krešimir Romić[1], and Danilo Babin[2]

[1] Faculty of Electrical Engineering, Computer Science and Information Technology Osijek, Josip Juraj Strossmayer University of Osijek, Osijek, Croatia
marija.habijan@ferit.hr

[2] TELIN-IPI, Faculty of Engineering and Architecture, Ghent University – imec, Ghent, Belgium

Abstract. Automatic cardiac MRI segmentation, including left and right ventricular endocardium and epicardium, has an essential role in clinical diagnosis by providing crucial information about cardiac function. Determining heart chamber properties, such as volume or ejection fraction, directly relies on their accurate segmentation. In this work, we propose a new automatic method for the segmentation of myocardium, left, and right ventricles from MRI images. We introduce a new architecture that incorporates SERes blocks into 3D U-net architecture (3D SERes-U-Net). The SERes blocks incorporate squeeze-and-excitation operations into residual learning. The adaptive feature recalibration ability of squeeze-and-excitation operations boosts the network's representational power while feature reuse utilizes effective learning of the features, which improves segmentation performance. We evaluate the proposed method on the testing dataset of the MICCAI Automated Cardiac Diagnosis Challenge (ACDC) dataset and obtain highly comparable results to the state-of-the-art methods.

Keywords: Cardiac MRI segmentation · Left ventricle · Right ventricle · Myocardium · Residual learning · Squeeze and excitation · 3D SERes-U-Net

1 Introduction

Cardiovascular diseases (CVDs) cause major health complications that often lead to death [19]. An evaluation of cardiac function and morphology plays an essential role for CVDs' early diagnosis, risk evaluation, prognosis setting, and therapy decisions. Magnetic resonance imaging (MRI) has a high resolution, contrast and great capacity for differentiating between types of tissues. This makes MRI the gold standard of cardiac function analysis [2]. Delineations of the myocardium (Myo), left ventricle (LV), and right ventricle (RV) are necessary for quantitative

G. Rozinaj and R. Vargic (Eds.): IWSSIP 2021, CCIS 1527, pp. 3–14, 2022.
https://doi.org/10.1007/978-3-030-96878-6_1

assessment and calculation of clinical indicators such as volumetric measures at end-systole and at end-diastole, ejection fraction, thickening measures, as well as mass. Semi-automatic delineation is still commonly present in clinical practice. That is often a laborious, and time-consuming process, prone to intra- and inter-observer variability. Hence, accurate, reliable, and automated segmentation methods are required to facilitate cardiovascular disease diagnosis.

Various image processing methods have been proposed to automatize segmentation tasks in the medical field [4,10,21]. While some of these approaches use more traditional techniques like level sets [17], registration and atlases [5,8], fully-automatic methods mostly employ fully convolutional neural networks (FCNNs) [6]. Commonly used approaches include structures that consist of a series of convolutional, pooling, and deconvolutional layers such as U-Net architecture [7,22]. Generally, various deep learning methods have shown outstanding performance on medical images for segmentation purposes [3,9,13–16,20,23,24]. Promising as they are, the appearance of overfitting on limited training data, vanishing and exploding gradients, and network degradation are significant concerns for FCNs. The residual learning, introduced in ResNets [11], overcomes the above problems by enhancing information flow over through the network using identity shortcut connections. Squeeze and excitation operations, introduced in SeNets, [12] improve the network's representational power by modeling interdependencies of channel-wise features and by dynamically recalibrating them.

Motivated by previously described advancements, we propose a 3D U-Net-based network that incorporates residual and squeeze and excitation blocks (SERes blocks). We introduce the squeeze and excitation (SE) blocks at 3D U-Nets' encoder and decoder paths after each residual block. We provide experimental results of the proposed network for the task of LV, RV, and Myo segmentation and show that our proposed approach obtains highly comparable results to the state-of-the-art.

2 Method

2.1 Squeeze and Excitation Residual Block

The SERes block takes the advantages of the squeeze and excitation operations [12] for adaptive feature recalibration and residual learning for feature reuse [11].

The 3D SERes block can be expressed with the following expression:

$$\mathbf{X}^{res} = F_{res}(\mathbf{X}) \tag{1}$$

where $\mathbf{X}$ refers to the input feature, $\mathbf{X}^{res}$ is the residual feature, and $F_{res}(\mathbf{X})$ is residual mapping that needs to be learned. The squeeze function which groups channel-wise statistics and global spatial information using global average pooling can be expressed with:

$$F_{sq}(\mathbf{x}_n^{res}) = p_n = \frac{1}{L \times H \times W} \sum_{i=1}^{L} \sum_{j=1}^{H} \sum_{k=1}^{W} x_n^{res}(i, j, k) \tag{2}$$

where $\mathbf{p} = [p_1, p_2, ..., p_n]$ and p_n is the $n-th$ element of $\mathbf{p} \in R^n$, where, $L \times H \times W$ is the spatial dimension of $\mathbf{F}^{res}$, $x_n^{res} \in R^{L \times H \times W}$ represents the feature map of the $n-th$ channel from the feature $\mathbf{X}^{res}$, and N referst to the residual mapping channels'. Scale values for the residual feature channels $\mathbf{s} \in R^N$ can be expressed with:

$$\mathbf{s} = F_{ex}(\mathbf{p},\mathbf{W}) = \sigma(\mathbf{W}_2\delta(\mathbf{W}_1\mathbf{p})) \tag{3}$$

where F_{ex} is the excitation function which generates them. It is parameterized by two fully connected layers (FCNs) with parameters $\mathbf{W}_1 \in R^{\frac{N}{r} \times N}$ and $\mathbf{W}_2 \in R^{N \times \frac{N}{r}}$, the sigmoid function σ, the $ReLU$ function δ and has reduction ration determened with r. The multiplication between feature map and learned scale value s_n across channel can be expressed with:

$$\widetilde{\mathbf{X}}_n^{res} = F_{scale}(\mathbf{X}_n^{res}, s_n) = s_n \cdot \mathbf{X}_n^{res}, \in R^{H \times W \times L} \tag{4}$$

Finally, applying the squeeze and excitation operations obtains the calibrated residual feature, which can be expressed with:

$$\widetilde{\mathbf{X}}^{res} = [\widetilde{\mathbf{X}}_1^{res}, \widetilde{\mathbf{X}}_2^{res}, ..., \widetilde{\mathbf{X}}_n^{res}] \tag{5}$$

The output feature $\mathbf{Y}$ after the $ReLU$ function δ is obtained as:

$$\mathbf{Y} = \delta(\widetilde{\mathbf{X}}^{res} + \mathbf{X}) \tag{6}$$

where $(\widetilde{\mathbf{X}}^{res} + \mathbf{X})$ refers to element-wise addition and a shortcut connection.

An illustration of the 3D ResNet block and 3D SERes block is shown in Fig. 1.

2.2 3D SERes-U-Net Architecture

Our proposed network architecture is based on the standard 3D U-Net [7] which follows encoder-decoder architecture. The encoder or contracting pathway encodes the input image and learns low-level features, while the decoder or expanding pathway learns high-level features and gradually recovers original image resolution.

Like 3D U-Net, our contracting pathway consist of three downsampling layers. We replace initially used pooling layers in the original 3D U-Net with convolutional layers with stride equal to 2. Instead of plain units, we adopt SERes blocks consisting of squeeze and excitation operations followed by a residual block, as described in 2.1, to accelerate convergence and training. Each residual block inside the SERes block has two convolutional layers that are followed by ReLU activation, and batch normalization (BN) as shown in Fig. 1(b). Similarly, three SERes blocks are used in the expanding pathway. This pathway has three up-sampling layers, each of which doubles the size of the feature maps, and are followed by a $2 \times 2 \times 2$ convolutional layer. The network can acquire the importance degree of each residual feature channel through the feature recalibration strategy. Based on the importance degree, the less useful channel features are

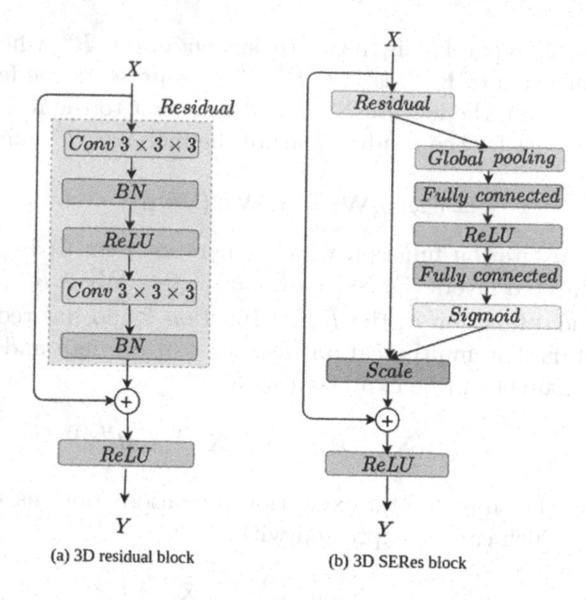

Fig. 1. An illustration of used residual blocks. (a) The original 3D ResNet block and (b) structure of the 3D SERes block

suppresed while useful features are enhanced. Therefore, by modeling the interdependencies between channels, the 3D SERes block performs dynamic recalibration of residual feature responses in a channel-wise manner. In this way, the network can capture every residual feature channel's importance degree, which improves its representational power. SERes-U-Net architecture is presented in Fig. 2.

3 Implementation Details

3.1 Dataset and Evaluation Metrics

The Automated Cardiac Diagnosis Challenge (ACDC) dataset consists of real-life clinical cases obtained from an everyday clinical setting at the University Hospital of Dijon (France). The dataset includes cine-MRI images of patients suffering from different pathologies, including myocardial infarction, hypertrophic cardiomyopathy, dilated cardiomyopathy, abnormal right ventricle, and normal cardiac anatomy. Dataset has been evenly divided based on the pathological condition and includes 100 cases with corresponding ground truth for training, and 50 cases for testing through an online evaluation platform. Clinical experts manually annotated LV, RV, and Myo at systolic and diastolic phases, for which the weight and height information was provided as well. Images are acquired as a series of short-axis slices covering the LV from the base to the apex. The spatial resolution goes from 1.37 to 1.68 mm^2/pixel, slice thickness is between 5–8 mm, while an inter-slice gap is 5 or 10 mm.

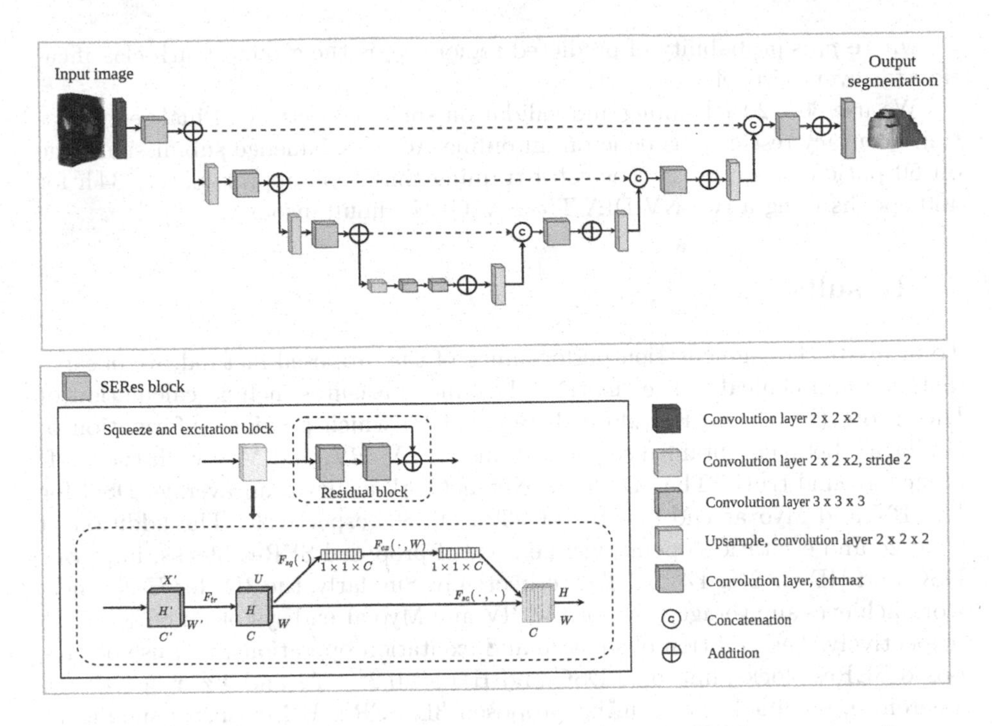

Fig. 2. Illustration of SERes-U-Net architecture for LV, RV, Myo segmentation.

3.2 Network Training

To overcome high intensity irregularities of MRI images, we normalize each volume based on the standard and mean deviation of their intensity values. The volumes were center-cropped to a fixed-size and zero-padded to provide fine ROI for the network input. For data augmentation, we apply random axis mirror flip with a probability of 0.5, random scale, and intensity shift on input image channel. We use $L2$ norm regularization with a weight of 10^{-5} and employ the spatial dropout with a rate of 0.2 after the initial encoder convolution. We use Adam optimizer with initial learning rate of $\alpha_0 = 10^{-4}$ and gradually decrease it according to following expression:

$$\alpha = \alpha_0 * \left(1 - \frac{e}{T_e}\right)^{0.9} \tag{7}$$

where T_e is a total number of epochs and e is an epoch counter. We employ a smoothed negative Dice score [18] loss function, defined with:

$$D_{loss} = -\frac{2\sum_{i=1}^{N} p_i g_i + 1}{\sum_{i=1}^{N} p_i + \sum_{i=1}^{N} g_i + 1} \tag{8}$$

where p_i is probability of predicted regions, g_i is the ground truth classification for every i voxel.

We use 80%-20% training and validation split, respectively. Final segmentation accuracy testing was done on an online ACDC Challenge submission page on 50 patient subjects [1]. The total training time took approximately 34 h for 400 epochs using a two NVIDIA Titan V GPU, simultaneously.

4 Results

To evaluate the segmentation performance of the proposed method, we observe distance and clinical indices metrics. Distance measures include calculation of Dice score (DSC) and Hausdorff distance (HD) which provides information of similarity between obtained segmentations for LV, RV, and Myo with their reference ground truth. The 3D Res-U-Net network achieves an average DSC for LV, RV and Myo at end diastole of 93%, 86, 80 respectively. The addition of squeeze and excitation operations, i.e., use of proposed SERes blocks, improves DSC and HD for 2%, 4% and 3%, respectively. Similarly, the 3D Res-U-Net network achieves an average DSC for LV, RV and Myo at end systole of 86%, 77, 81 respectively. The addition of squeeze and excitation operations, i.e., use of proposed SERes blocks, improves DSC and HD for 0.2%, 6% and 4%, respectively. Therefore, obtained results using proposed 3D SERes-U-Net shows significant improvements in DSC in comparison to network without squeeze and excitation operations (3D Res-U-Net). Detailed qualitative results are shown in Table 1 and Table 2 while Fig. 3 provides visual example of obtained segmentation predictions. Clinical metrics include calculation of the most widely used indicators of hearts' function, including volume of the left ventricle at end-diastole (LVEDV), volume of the left ventricle at end-systole (LVESV), left ventricles' ejection fraction (LVEF), volume of the right ventricle at end-diastole (RVEDV), volume of the right ventricle at end-systole (RVESV), right ventricles' ejection fraction (RVEF), myocardium volume at end-systole (MyoLVES), and myocardium mass at end-diastole (MyoMED).

Table 1. The segmentation accuracy results for LV, RV and Myo expressed in Dice score (DSC) and Hausdorff distance (HD) for the proposed method at end diastole for 3D Res-U-Net and proposed 3D SERes-U-Net.

Methods	ED					
	LV		RV		Myo	
	D_{sc}	H_d	D_{sc}	H_d	D_{sc}	H_d
3D Res-U-Net	0.93 (0.0636)	38.2 (4.8721)	0.86 (0.0919)	52.9 (12.4414)	0.8 (0.0636)	32.95 (5.6003)
3D SERes-U-Net	0.95 (0.0071)	11.53 (0.4101)	0.9 (0.0212)	23.41 (12.3571)	0.83 (0.0071)	13.77 (1.9871)

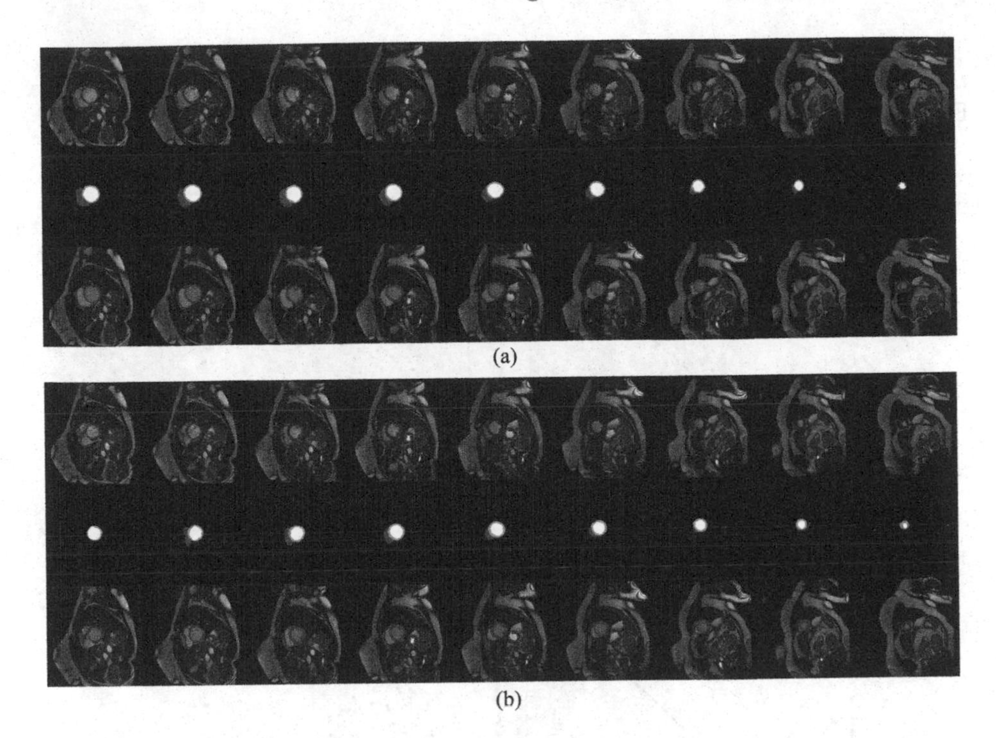

Fig. 3. An example of obtained results. (a) Top row: original MRI image at end diastolic phase of cardiac cycle. Middle row: Obtained segmentation. Bottom row: an overlay of original image and obtained segmentation prediction. (b) Top row: original MRI image at end systolic phase of cardiac cycle. Middle row: Obtained segmentation. Bottom row: an overlay of original image and obtained segmentation prediction.

Table 2. The segmentation accuracy results for LV, RV and Myo expressed in Dice score (DSC) and Hausdorff distance (HD) for the proposed method at end systole for 3D Res-U-Net and proposed 3D SERes-U-Net.

Methods	ES					
	LV		RV		Myo	
	D_{sc}	H_d	D_{sc}	H_d	D_{sc}	H_d
3D Res-U-Net	0.86 (0.0283)	29.77 (1.7748)	0.77 (0.0424)	36.99 (5.3952)	0.81 (0.0283)	30.29 (1.1031)
3D SERes-U-Net	0.86 (0.1273)	11.94 (8.4994)	0.83 (0.0283)	21.49 (5.7558)	0.85 (0.0071)	15.00 (1.9799)

The Pearson correlation coefficient (R) and Bland-Altman and analysis of the results obtained using proposed methed for LV, RV and Myo are shown in Figs. 5, 6, 4.

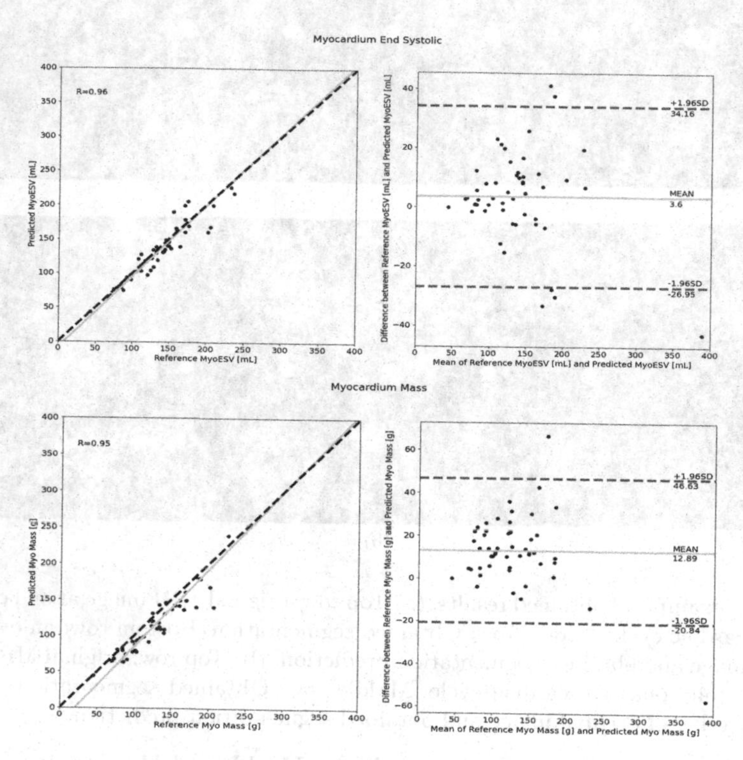

Fig. 4. Comparison of the automatically obtained segmentations and the reference volume of the myocardium end systolic volume and myocardium mass. The image showns correlation and Bland-Altman plots to compare automatically obtained segmentation and the reference values.

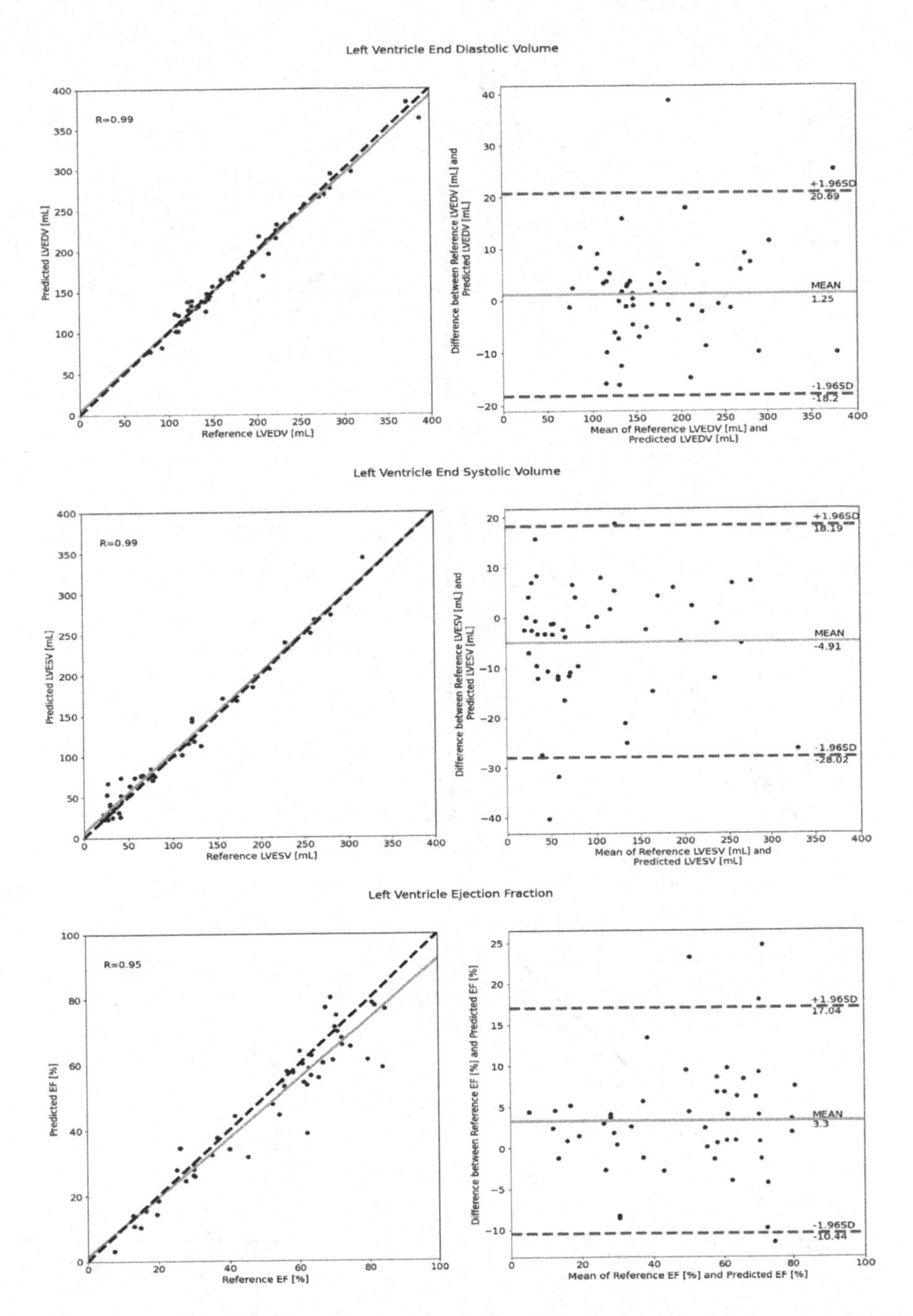

Fig. 5. Comparison of the automatically obtained segmentations and the reference volumes of the MRI scans. The image showns correlation and Bland-Altman plots for the LV volumes at and diastole and at the end systole as well as ejection fraction.

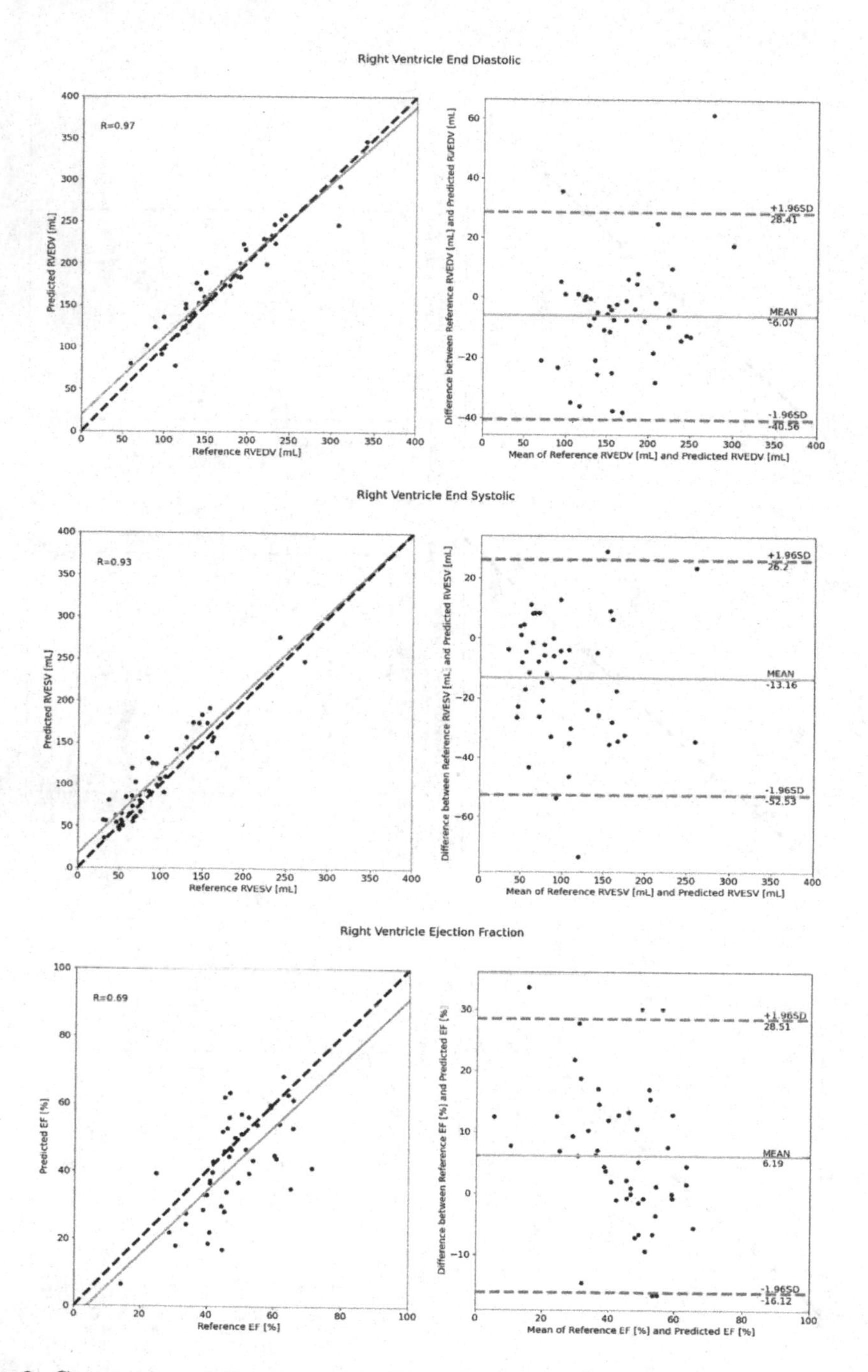

Fig. 6. Comparison of the automatically obtained segmentations and the reference volumes of the MRI scans. The image showns correlation and Bland-Altman plots for the RV volumes at and diastole and at the end systole as well as ejection fraction.

5 Conclusion

In this work, a deep neural network architecture named 3D SERes-U-Net was introduced for automatic segmentation of LV, RV, and Myo from MRI images. The significance of the proposed approach is in the two main characteristics. First, the approach is based on 3D deep neural networks, which are suitable for volumetric medical image processing. Second, the network introduces SERes blocks which optimizes the deep network and extracts distinct features. By taking advantage of the 3DSERes block, the proposed method learns the features with high discrimination capability, which is favorable to identify cardiac structures from the complex environment.

Acknowledgement. This work has been supported in part by Croatian Science Foundation under the Project UIP-2017-05-4968.

References

1. A.C.D.C.A.M.C.: Post-2017-miccai-challenge testing phase. https://acdc.creatis.insa-lyon.fr/#challenges (2017)
2. Arnold, J.R., McCann, G.P.: Cardiovascular magnetic resonance: applications and practical considerations for the general cardiologist. Heart **106**(3), 174–181 (2020)
3. Baumgartner, C.F., Koch, L., Pollefeys, M., Konukoglu, E.: An exploration of 2d and 3d deep learning techniques for cardiac MR image segmentation. ArXiv arXiv:1709.04496 (2017)
4. Bernard, O., et al.: Deep learning techniques for automatic MRI cardiac multi-structures segmentation and diagnosis: is the problem solved? IEEE Trans. Med. Imaging **37**(11), 2514–2525 (2018). https://doi.org/10.1109/TMI.2018.2837502
5. Cetin, I., et al.: A radiomics approach to computer-aided diagnosis with cardiac cine-MRI. In: Pop, M., et al. (eds.) STACOM 2017. LNCS, vol. 10663, pp. 82–90. Springer, Cham (2018). https://doi.org/10.1007/978-3-319-75541-0_9
6. Cheng, F., et al.: Learning directional feature maps for cardiac MRI segmentation. In: Martel, A.L., et al. (eds.) MICCAI 2020. LNCS, vol. 12264, pp. 108–117. Springer, Cham (2020). https://doi.org/10.1007/978-3-030-59719-1_11
7. Click, O., Abdulkadir, A., Lienkamp, S.S., Brox, T., Ronneberger, O.: 3d u-net: Learning dense volumetric segmentation from sparse annotation. CoRR abs/1606.06650 (2016). http://arxiv.org/abs/1606.06650
8. Duan, J., et al.: Automatic 3d bi-ventricular segmentation of cardiac images by a shape-refined multi- task deep learning approach. IEEE Trans. Med. Imaging **38**(9), 2151–2164 (2019)
9. Habijan, M., Leventić, H., Galić, I., Babin, D.: Estimation of the left ventricle volume using semantic segmentation. In: 2019 International Symposium ELMAR, pp. 39–44 (2019). https://doi.org/10.1109/ELMAR.2019.8918851
10. Habijan, M., et al.: Overview of the whole heart and heart chamber segmentation methods. Cardiovasc. Eng. Technol. **11**(6), 725–747 (2020)
11. He, K., Zhang, X., Ren, S., Sun, J.: Deep residual learning for image recognition. In: 2016 IEEE Conference on Computer Vision and Pattern Recognition (CVPR), pp. 770–778 (2016). https://doi.org/10.1109/CVPR.2016.90

12. Hu, J., Shen, L., Sun, G.: Squeeze-and-excitation networks. In: 2018 IEEE/CVF Conference on Computer Vision and Pattern Recognition, pp. 7132–7141 (2018). https://doi.org/10.1109/CVPR.2018.00745
13. Isensee, F., Jaeger, P., Full, P.M., Wolf, I., Engelhardt, S., Maier-Hein, K.H.: Automatic cardiac disease assessment on cine-MRI via time-series segmentation and domain specific features. CoRR abs/1707.00587 (2017). http://arxiv.org/abs/1707.00587
14. Jang, Y., Hong, Y., Ha, S., Kim, S., Chang, H.-J.: Automatic segmentation of LV and RV in cardiac MRI. In: Pop, M., et al. (eds.) STACOM 2017. LNCS, vol. 10663, pp. 161–169. Springer, Cham (2018). https://doi.org/10.1007/978-3-319-75541-0_17
15. Khened, M., Varghese, A., Krishnamurthi, G.: Densely connected fully convolutional network for short-axis cardiac cine MR image segmentation and heart diagnosis using random forest. In: STACOM@MICCAI (2017)
16. Liu, T., Tian, Y., Zhao, S., Huang, X., Wang, Q.: Residual convolutional neural network for cardiac image segmentation and heart disease diagnosis. IEEE Access **8**, 82153–82161 (2020). https://doi.org/10.1109/ACCESS.2020.2991424
17. Liu, Y., et al.: Distance regularized two level sets for segmentation of left and right ventricles from cine-MRI. Magn. Reson. Imaging **34** (2015). https://doi.org/10.1016/j.mri.2015.12.027
18. Lu, J.-T., et al.: DeepAAA: clinically applicable and generalizable detection of abdominal aortic aneurysm using deep learning. In: Shen, D., et al. (eds.) MICCAI 2019. LNCS, vol. 11765, pp. 723–731. Springer, Cham (2019). https://doi.org/10.1007/978-3-030-32245-8_80
19. Organization, W.H.: Mortality database (2018). Accessed 19 Jan 2021
20. Patravali, J., Jain, S., Chilamkurthy, S.: 2d–3d fully convolutional neural networks for cardiac MR segmentation. ArXiv arXiv:1707.09813 (2017)
21. Peng, P., Lekadir, K., Gooya, A., Shao, L., Petersen, S.E., Frangi, A.F.: A review of heart chamber segmentation for structural and functional analysis using cardiac magnetic resonance imaging. Magn. Reson. Mater. Phys. Biol. Med. **29**(2), 155–195 (2016). https://doi.org/10.1007/s10334-015-0521-4
22. Ronneberger, O., Fischer, P., Brox, T.: U-net: convolutional networks for biomedical image segmentation. CoRR abs/1505.04597 (2015). http://arxiv.org/abs/1505.04597
23. Vigneault, D.M., Xie, W., Ho, C.Y., Bluemke, D.A., Noble, J.A.: Omega-net (omega-net): fully automatic, multi-view cardiac MR detection, orientation, and segmentation with deep neural networks. Med. Image Anal. **48**, 95–106 (2018). https://doi.org/10.1016/j.media.2018.05.008
24. Zotti, C., Luo, Z., Humbert, O., Lalande, A., Jodoin, P.M.: Gridnet with automatic shape prior registration for automatic MRI cardiac segmentation. In: STACOM@MICCAI (2017)

Fingerprint Classification Based on the Henry System via ResNet

João W. Mendes de Souza[1,3(✉)], Aldisio G. Medeiros[1,3], Gabriel B. Holanda[1,2], Paulo A. L. Rego[3], and Pedro P. Rebouças Filho[1,2]

[1] Image Processing Laboratory, Signs and Applied Computer (LAPISCO), Fortaleza, Brazil
{wellmendes,aldisio.medeiros,gabrielbandeira}@lapisco.ifce.edu.br
[2] Federal Institute of Ceará (IFCE), Fortaleza, Ceará, Brazil
pedrosarf@ifce.edu.br
[3] Federal University of Cear (UFC), Fortaleza, Ceará, Brazil
pauloalr@ufc.br

Abstract. Fingerprints are widely used for biometric validation worldwide. Since human fingerprints are unique and remain constant over time, it provides an easy-to-use, reliable, and economical authentication method. In addition to that, fingerprint recognition systems are of great importance because of their applicability in our lives. This work presents a classification methodology based on Henry Classification System using Convolutional Neural Networks (CNNs) models, such as Darknet, Alexnet, Resnet, VGG16, and Deep Belief Network. Besides that, we evaluate our proposal by carrying out experiments using grayscale images and pre-processed images as input on the classification step with the combination of the Gabor filter and the morphological thinning operation. We have obtained the highest result accuracy of 95.1% in the NIST Special Database 4, a widespread fingerprint dataset, using the Resnet 34 model in grayscale images. The proposed approach was evaluated with extraction strategies of classic attributes and based on convolutional networks. According to the results, the proposed methodology presents promising results, surpassing the traditional approaches present in the literature.

Keywords: Fingerprint · Henry classification system · Classification · CNNs

1 Introduction

Fingerprints are impressions left on the surfaces by the friction ridges on a human's finger. The friction ridge refers to the upper epidermis' elevated part and is composed of connected friction ridge skin units.

Currently, fingerprints are the most popular biometric features used for personal identification. It presents an easy-to-use, reliable, and economical way

G. Rozinaj and R. Vargic (Eds.): IWSSIP 2021, CCIS 1527, pp. 15–28, 2022.
https://doi.org/10.1007/978-3-030-96878-6_2

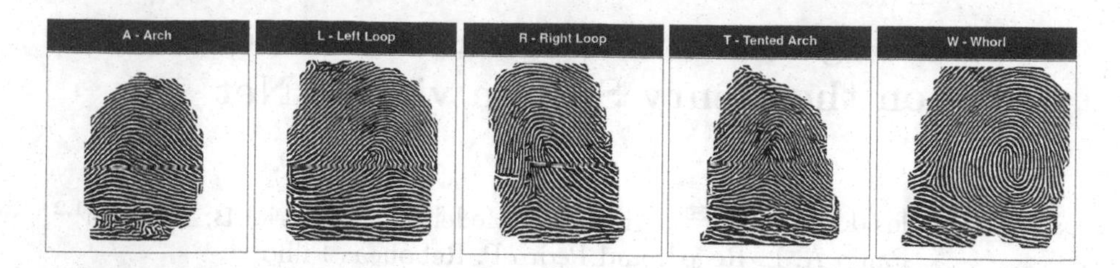

Fig. 1. Example of each class in NIST special database 4 based on henry classification system [12] after pre-processing.

to authenticate an individual since a human fingerprint is unique and remains invariant over time [30].

Thereby, fingerprint prediction systems are an essential tool that groups fingerprints according to their features and compares an input fingerprint with an extensive fingerprint database. The query fingerprints that need to be combined can be compared with a subset of fingerprints in the existing database [2,21].

The general structure of a fingerprint prediction system consists of four main steps [21]:

1. The acquisition of fingerprint features is the process of obtaining a scanned image of a person using a specific capture device.
2. Pre-processing allows improving the overall quality of the captured image.
3. Data features are extracted using different algorithms.
4. The classification of the extracted features is usually applied to perform the individual's recognition.

The performance of fingerprint-based systems depends directly on the reliability and precision of the feature extraction stage [29]. The matching algorithms are based on the pairing of features found in fingerprints. Abbood *et al.* [1] suggest that the reliability of the extracted resources is related to the fingerprint's quality. Schuch *et al.* [28] indicate that the application of image enhancement filters improves the reliability of the extracted resources.

The fingerprint classification problem has been widespread in the scientific community for a long time [14]. Thereby, the Henry Classification System was widely used, through the distribution in classes, which present their characteristics, some of them pertinent to the delta[1] and the other lines of the nuclear system [12].

The American National Standards Institute (ANSI) classified fingerprints based on the points of singularity proposed by Henry (Loop, Delta, and Whorl) [12], in five different classes: (I) arch; (II) tented arch; (III) whorl; (IV) right loop; (V) left loop. Examples of these classes are shown in Fig. 1.

This article presents a methodology for classifying fingerprints, using the Henry Classification System based approach, using Convolutional Neural Networks (CNNs) models. In addition, the objective is to make comparisons with

[1] A point in loop and whorl prints that lies within an often triangular, three-pronged, or funnel-shaped structure.

classic learning methods to assess the impact of promising technologies in the context of digital classification. Besides that, we compared the results obtained with those between the CNNs models. We will also evaluate the effects of filters for pre-processing fingerprint images before classification.

The work is organized as shown: Sect. 2 presents some related works. Section 3 presents the methodology proposed in this work. In Sect. 4, we show the results obtained with this methodology. Finally, we present the conclusion of the work in Sect. 5.

2 Related Works

This section will briefly present works found in the literature related to fingerprint classification and recognition in digital images using Convolutional Neural Networks (CNNs) and classical machine learning methods. Fingerprint classification techniques solve various daily problems.

Several authors have recently addressed the subject of fingerprint classification. Among them, we can mention Win *et al.* work's [35], which presented a review of fingerprint classification and identification methods in the context of criminal investigation. The proposed methodology of Almajmaie *et al.* [3] presented a digital recognition approach based on a modified multi-connection architecture (MMCA) that was applied to the international database NIST 4, which was also used in our work. Finally, works that are not so recent, such as that by Liu [22], which presents an approach using the Adaboost classifier, learning with singularity resources, can be compared with our work since it uses a classic learning method.

A curious work created by Chhabra *et al.* [5], try to segment and classify unintentional fingermarks left at crime scenes. The method uses ensemble techniques to search best-extracted attributes based on Random Decision Forest (RDF) and Adaboost classifier results.

Some works apply Deep Learning in fingerprint [7,15], such as the work of Rim *et al.* [27], which identifies detailed fingerprint information using the Deep learning approach. Their work developed private Dataset of fingerprints and reached the best case accuracy of 90.98%.

Le *et al.* [19] developed a novel algorithm for fingerprint image enhancement. Their methodology uses adaptive multi-linear algebra, higher-order SVD (HOSVD) on a tensor of wavelet subbands. In the best case, the proposed method achieved a classification accuracy of 98.05%.

3 Evaluation Methodology

As shown in Fig. 2, in this work, the evaluation methodology focuses on two studying lines. The first evaluates the input images without applying pre-processing operations (blue arrow). In this case, we want to assess the real need to apply some effort to improve the image. A second line (gray arrow) evaluates the images from the Gabor filter's enhancement, followed by the application of a

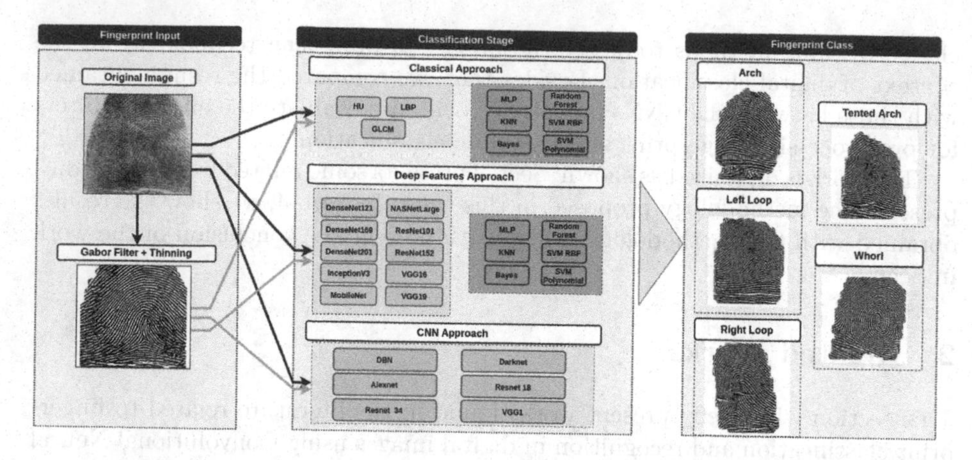

Fig. 2. Flowchart of the proposed evaluation methodology. The three blocks in the center of the image represent the different approaches evaluated in this work. (Color figure online)

thinning operation. It's worth mentioning that we choose Gabor filter's enhancement because it is widely used in fingerprint applications since its purpose is texture analysis.

Both experiments are evaluated in three different image classification approaches: The first scenario, called the Classical Approach, evaluates methods of extracting traditional attributes by combining them with traditional classifiers. The second scenario evaluates feature extractors' performance based on convolutional neural networks, producing results known as Deep Features. Finally, the last scenario assesses six recent convolutional networks for classifying incoming samples. As a way out of the system, we have a class prediction based on Henry's classification system. In all cases, the evaluation metrics used are Accuracy (ACC), Positive predictive value (PPV), Sensitivity (Sen), and F1-Score (F_1S).

3.1 NIST Special Database 4

The NIST Special Database 4, refered as "NIST 8-Bit Gray scale Images of Fingerprint Image Groups (FIGS)" [34], is a database composed of 8-bit grayscale images with a resolution of 512 × 512 pixels from fingerprint groups. This database contains 4000 images classified based on the Henry Classification System [12], divided equally into 800 pictures for each of the following classes: Arch (A), Left loop (L), Right loop (R), Tented arch (T), and Whorl (W) as shown in Fig. 1.

3.2 Automatic Fingerprint Classification

Pre-processing. The pre-processing adopted in this methodology consists of filtering combined with a thinning morphological operation applied in the fingerprint image as grayscale. This stage aims to obtain a fingerprint image with enhanced ridges, emphasizing the delta and core that are the main regions to the classification stage.

We use a Gabor filter for filtering, a linear filter used for texture analysis, which essentially means that it analyzes any specific frequency content in the image due to a region located around the point or in analyzed areas [8].

The Gabor filter is a Gaussian kernel function modulated by a complex sinusoidal plane wave [10], defined as:

$$G(x, y) = \frac{f^2}{\pi\gamma\eta} exp(-\frac{x'^2 + \gamma^2 y'^2}{2\sigma^2}) exp(j2\pi f x' + \phi) \tag{1}$$

where x'^2 and y'^2 are, respectively, $xcos\theta + ysin\theta$ and $-xsin\theta + ycos\theta$, f is the frequency of the sinusoid, θ represents the stripes orientation, ϕ is the phase offset, σ is the standard deviation of the Gaussian envelope and γ is the spatial aspect ratio for the ellipticity.

Finally, we use a morphological thinning operation, which transforms a digital image into a simplified, topologically equivalent image. It is a type of topological skeleton, but using mathematical morphology operators [18]. The morphological thinning operation is defined as [9]:

$$I \otimes K = I \cap (I \circledast K)^c \tag{2}$$

where I is the two-dimensional input image and K is a cross structuring element (kernel) of 5×5.

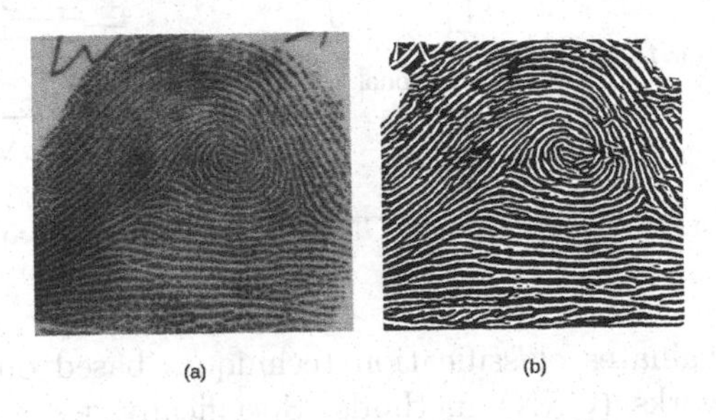

Fig. 3. Example of pre-processing using the proposed approach (a) original image (b) pre-processed image.

Figure 3 presents an example of pre-processing using a Gabor filter and then the morphological operation of thinning in an image of the dataset.

Fingerprint Classification. On the learning stage, aiming to perform the fingerprint classification based on the five different fingerprint classes. To improve our evaluation methodology, we apply classical machine learning methods, which use a combination of classical feature extractors. As feature extractors, we applied the Grey Level Co-occurrence Matriz (GLCM), Hu Moments, and Local

Binary Patterns (LBP), and as machine learning classification methods, we evaluated Naive Bayes, k-Nearest Neighbors (kNN), Multi-Layer Perceptron (MLP), Random-Forest (RF), and Support Vector Machine (SVM).

Unlike the traditional approach, CNNs have in their structure the ability to learn better convolutional filters for processing samples during the training process. This adaptive potential for identifying characteristics is explored in this work as a feature extractor. Previous work has applied this strategy to fingerprint images achieved promising results [24].

To transform CNNs into attribute extractors, we use pre-trained network structures to classify objects present in the ImageNet [17] dataset. We removed the fully connected layer from the CNN structure that estimates a percentage probability by class present in the dataset. At the network output, we have a vector of attributes that were produced throughout the entire architecture, from the input layer until the penultimate layer. So, we have an attribute extractor instead of a classifier. Figure 4 illustrates the vectorization of the last convolutional layer to form a single vector of attributes. In this proposal, we will evaluate ten different recent CNN architectures used in Deep Features approach [24]: DenseNet121, DenseNet169, DenseNet201, InceptionV3, MobileNet, NASNetLarge, ResNet101, ResNet152, VGG16, and VGG19.

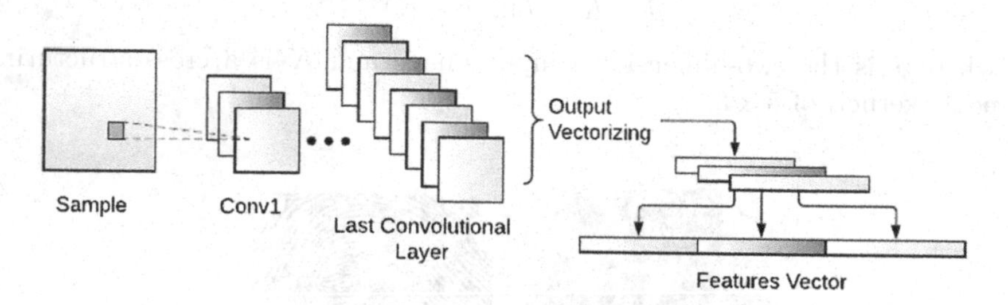

Fig. 4. Deep feature extraction from CNN Architectures. Adapted from [24].

Present work evaluates classification techniques based on recent Convolutional Neural Networks (CNN) methods. Specifically, we assessed the open-source framework called Darknet. The Darknet is an open source neural network framework written in C and CUDA language [26]. In this work, we evaluated six different CNN models from scratch for the classification stage. The first one is a straightforward and well-known model called a two-layer convolutional Deep Belief Network (DBN) [16], with input image size of 28 $\times$ 28 pixels. The second model used was the Darknet [31], with input image size of 256 $\times$ 256 pixels. The third model used was the Alexnet [17], with input image size of 257 $\times$ 257 pixels. The fourth, fifth, and sixth are Resnet 18 and Resnet 34 [11], VGG16 [23], respectively, all this models using input image size of 256 $\times$ 256 pixels.

4 Results

In this section, we evaluate the proposed methodology by comparing classic learning methods using CNNs for image classification. This experiment uses NIST Special Database 4, which contains 4000 images of 512 × 512 pixels.

4.1 Classical Approach

Table 1. Fingerprint classification using classic methods with grayscale images and pre-processed images

Extractor	Model	Grayscale (%)				Gabor + Thinning (%)			
		Acc	PPV	Sen	F_1S	Acc	PPV	Sen	F_1S
GLCM	Bayes	24.0	13.7	24.0	14.2	20.0	4.0	20.0	6.7
	MLP	26.9	22.1	26.9	19.8	31.4	23.2	31.4	25.3
	KNN	24.6	24.2	24.6	24.0	31.0	29.1	31.0	29.0
	RF	27.2	26.1	27.2	24.4	33.7	30.2	33.7	28.5
	SVM (Polynomial)	22.9	18.8	22.9	11.6	35.1	33.5	35.1	29.5
	SVM (RBF)	26.9	16.2	26.9	18.1	31.7	24.4	31.7	25.3
HU	Bayes	22.0	18.1	22.0	17.9	19.3	13.4	19.3	15.4
	MLP	23.1	23.9	23.1	21.7	19.0	14.9	19.0	16.2
	KNN	20.7	20.7	20.7	20.7	21.3	21.6	21.3	20.1
	RF	20.3	13.0	20.3	9.6	21.1	20.3	21.1	17.3
	SVM (Polynomial)	23.2	24.3	23.2	21.1	20.5	24.6	20.5	15.3
	SVM (RBF)	21.6	20.5	21.6	19.9	20.2	19.0	20.2	18.3
LBP	Bayes	26.4	24.0	26.4	21.4	41.1	39.6	41.1	37.4
	MLP	27.5	20.6	27.5	21.3	35.9	35.7	35.9	29.9
	KNN	20.1	12.6	20.1	9.8	36.8	39.4	36.8	33.6
	RF	19.2	11.2	19.2	10.0	45.3	44.2	45.3	43.7
	SVM (Polynomial)	21.2	9.8	21.2	10.8	24.7	18.5	24.7	13.9
	SVM (RBF)	20.9	26.3	20.9	16.2	37.3	38.6	37.3	32.2

In order to assess the approach we executed 100 independent executions having the dataset shuffled and then splitted in 75% for training with Cross Validation, and 25% for testing.

According to Henry's classification system, we present in Table 1 the classification results using a classic approach for the five fingerprint classes. Table 1 shows the results for the classification of images from fingerprints in grayscale, without any enhancement pre-processing. We can observe that the HU extractor presented the worst results, reaching a maximum of 23% accuracy on average. On the other hand, the LBP extractor showed the classical approach's best performance, coming at 27.5% accuracy when combined with the MLP classifier. These results, however, are well below desirable. The classical approach's performance indicates that the evaluated feature extractors do not have a robust potential for identifying characteristics, capable of discriminating samples from a single class, failing in approximately 70% of cases, on average.

Table 1 also presents pre-processed images with a Gabor filter to enhance papillary ridges, followed by a morphological thinning process. As we can see, it is possible to notice an improvement in the classic metrics. The best combination is presented by the LBP extractor combined with Random Forest, reaching 45.3%. In addition to being better than the results using grayscale images, the results with pre-processed images indicate that the applied enhancement favored the process of class separation, assisting attribute extractors in identifying better characteristics. However, the classic approach could not learn enough features to correctly identify the samples, failing in more than 50% of cases, on average.

Table 2. Fingerprint classification using convolutional neural networks

Pre-processing	Model	Acc (%)	PPV (%)	Sen (%)	F1-Score (%)
Grayscale	DBN	60.2	34.7	36.2	31.4
	Darknet	93.9	85.9	85.5	85.6
	Alexnet	91.3	81.5	80.2	80.4
	Resnet 18	92.6	83.7	82.8	82.9
	Resnet 34	**95.1**	**88.7**	**88.4**	**88.4**
	VGG16	63.3	35.2	39.8	34.5
Gabor+Thin.	DBN	58.2	31.7	34.2	27.6
	Darknet	93.3	87.0	84.1	83.9
	Alexnet	93.4	84.7	84.5	84.5
	Resnet 18	94.3	89.3	86.9	86.9
	Resnet 34	**94.6**	**88.3**	**87.0**	**87.1**
	VGG16	70.5	50.9	46.6	43.5

4.2 Deep Features Approach

As a more recent approach, Table 4 presents the results using a classification based on the extraction of attributes through deep neural networks. In this experiment, we evaluated ten different CNN architectures. Each of the CNNs was used only as attribute extractors and then combined with traditional classification algorithms.

Table 4 presents the classification results for digital printing images in grayscale, without any preprocessing. As for accuracy, it is possible to observe that CNN InceptionV3 reaches lower metrics, with 23.3% when combined with the SVM classifier of the polynomial type. The MLP classifier presented the most stable results in terms of hit rates, showing superior results, on average, considering all combinations of attribute extractors. On the other hand, CNN

MobileNet combined with SVM type RBF the best performance, reaching 70.1% correct classifications.

Considering the fingerprint enhanced images with the Gabor filter followed by the application of morphological thinning, it is also possible to observe in Table 4 that the best combination of CNN as a feature extractor was DenseNet201 with MLP, reaching 78.1% accuracy. This result surpasses up to 32.8%, the best index presented by the classic approach with LBP and Random Forest. Finally, it is possible to conclude that the feature extraction approach using CNN presents an information extraction capacity that helps the classifiers better discriminate the samples in the real classes. It is believed that this result is achieved by the adaptive potential in the selection of convolutional filters proper to the learning strategy of convolutional networks. Thus CNN gives a greater representation capacity than in classical extraction.

4.3 CNN Approach

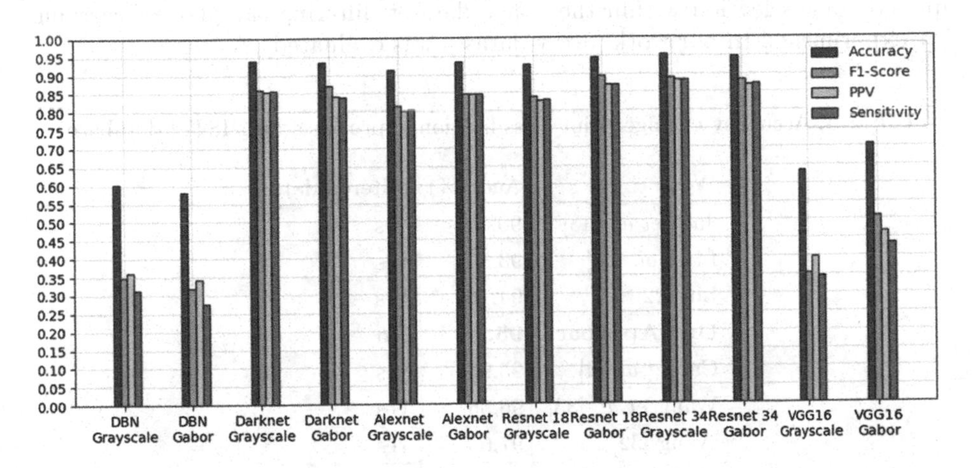

Fig. 5. Comparison with a bar plot between fingerprint classification results using CNNs.

The dataset was randomly divided into training and testing, where 25% of the data was destined for testing.

Performing analysis on the results obtained, we can see that the best result using a classic approach with grayscale images was 27.5% accurate, combining the LBP extractor with the MLP classifier. On the other hand, using classic methods with Gabor filter + Thinning, there was a gain of 8.4% accuracy, and the best result obtained with this approach was 45.3%.

Analyzing Table 2 and Fig. 5, which presents the approach using CNNs, we can see that the worst result was an accuracy of 60.2% using the DBN model with grayscale images. On the other hand, the best result showed an accuracy of 95.1% using the Resnet 34 model with images in grayscale as well. The results obtained comprise a significant difference between the approaches since there was a gain of 49.8% of the best result using the classical methods for the best result using CNNs.

It is worth mentioning that the pre-processing showed significant gains for classic approaches; on the other hand, it presented losses for the process using CNNs. Based on the F1-Score, we can emphasize that there was no discrepancy in the dataset used since the presence of false positives and negatives are always in the normal range.

Finally, our best approach obtained in the experiments, achieved accuracy of 95.1%. We emphasize that the proposed approach presented an accuracy on satisfactory scales and competitive potential compared to other literature approaches. Table 3 shows a comparison between different methodologies for classifying fingerprints within the Henry system. However, we emphasize that all strategies applied some selection within the image dataset, filtering samples for carrying out experiments. In our work, every dataset was evaluated.

Table 3. Accuracy of fingerprint classification approaches on NIST-4 database

Work	Acc. (%)	Filtered data
Jain *et al.* [13]	90.00	Yes
Li *et al.* [20]	93.50	Yes
Liu [22]	94.10	Yes
Our Approach	**95.10**	**No**
Cao *et al.* [4]	95.90	Yes
Wang *et al.* [33]	96.55	Yes
Wang [32]	97.65	Yes
Miranda [25]	95.05	Yes
Ding [6]	93.83	Yes

Table 4. Fingerprint classification using Deep Features combined with traditional classification methods with grayscale images and pre-processed images

Extractor	Model	Grayscale (%)				Gabor + Thinning (%)			
		Acc	PPV	Sen	F_1S	Acc	PPV	Sen	F_1S
DenseNet121	Bayes	47.8	47.1	47.8	46.8	57.4	58.6	57.4	57.3
	MLP	60.4	61.0	60.4	60.5	71.1	71.2	71.1	71.0
	KNN	49.4	49.3	49.4	48.8	59.5	59.8	59.5	59.4
	RF	51.9	53.8	51.9	49.7	67.3	68.9	67.3	67.1
	SVM (Poly)	38.8	48.4	38.8	33.1	46.3	52.9	46.3	44.7
	SVM (RBF)	60.6	60.8	60.6	59.8	70.3	71.0	70.3	70.3
DenseNet169	Bayes	49.8	49.0	49.8	48.9	62.7	63.0	62.7	62.1
	MLP	62.6	62.9	62.6	62.7	75.2	75.6	75.2	75.3
	KNN	53.7	53.3	53.7	52.3	66.2	66.8	66.2	66.1
	RF	48.8	51.2	48.8	45.5	71.2	72.7	71.2	70.6
	SVM (Poly)	47.8	50.8	47.8	41.1	51.9	60.7	51.9	50.5
	SVM (RBF)	66.0	65.8	66.0	65.0	77.4	77.9	77.4	77.3
DenseNet201	Bayes	52.8	52.8	52.8	52.0	62.5	65.0	62.5	62.4
	MLP	63.9	64.1	63.9	63.9	**78.1**	**78.6**	**78.1**	**78.2**
	KNN	54.3	55.2	54.3	54.2	66.8	66.9	66.8	66.4
	RF	45.3	51.3	45.3	41.1	73.9	75.3	73.9	73.9
	SVM (Poly)	47.6	61.4	47.6	46.1	42.3	52.3	42.3	36.5
	SVM (RBF)	65.6	67.1	65.6	66.0	77.5	77.9	77.5	77.5
InceptionV3	Bayes	40.0	37.8	40.0	38.1	48.8	48.1	48.8	48.2
	MLP	54.2	54.2	54.2	54.1	63.3	63.4	63.3	63.3
	KNN	43.0	41.6	43.0	41.7	50.7	49.8	50.7	49.9
	RF	41.3	43.0	41.3	38.0	56.3	55.5	56.3	54.4
	SVM (Poly)	23.3	16.7	23.3	12.3	26.8	26.0	26.8	18.5
	SVM (RBF)	52.5	51.3	52.5	50.4	65.1	64.8	65.1	64.8
MobileNet	Bayes	51.5	52.4	51.5	50.8	56.1	56.6	56.1	55.2
	MLP	66.5	66.6	66.5	66.4	68.6	69.0	68.6	68.7
	KNN	54.6	54.6	54.6	53.9	63.1	62.8	63.1	62.7
	RF	58.7	58.6	58.7	57.7	67.8	68.5	67.8	67.9
	SVM (Poly)	68.3	68.9	68.3	68.2	70.9	71.6	70.9	70.9
	SVM (RBF)	**70.1**	**71.0**	**70.1**	**70.3**	72.4	73.0	72.4	72.4
NASNetLarge	Bayes	44.9	43.4	44.9	43.0	43.6	45.3	43.6	43.4
	MLP	59.0	59.5	59.0	59.1	64.9	64.6	64.9	64.7
	KNN	47.8	48.1	47.8	46.6	50.4	52.8	50.4	50.3
	RF	50.8	50.5	50.8	49.6	57.1	57.4	57.1	56.5
	SVM (Poly)	34.5	30.7	34.5	22.6	36.2	47.6	36.2	29.4
	SVM (RBF)	58.4	59.5	58.4	58.2	64.4	65.3	64.4	64.5
ResNet101	Bayes	30.0	30.4	30.0	27.7	49.4	49.5	49.4	47.5
	MLP	46.6	47.0	46.6	46.7	71.8	72.0	71.8	71.8
	KNN	31.6	31.1	31.6	31.1	58.9	58.6	58.9	58.3
	RF	39.2	38.5	39.2	37.3	66.8	67.7	66.8	66.6
	SVM (Poly)	28.1	24.3	28.1	22.0	36.9	26.0	36.9	28.3
	SVM (RBF)	44.9	44.5	44.9	44.0	73.6	73.6	73.6	73.5
ResNet152	Bayes	30.6	31.3	30.6	28.7	52.9	51.7	52.9	50.9
	MLP	43.8	44.3	43.8	43.9	75.4	76.0	75.4	75.7
	KNN	30.1	31.3	30.1	29.8	62.6	62.9	62.6	62.5
	RF	33.9	35.8	33.9	31.4	70.8	71.0	70.8	70.5
	SVM (Poly)	26.0	29.4	26.0	19.5	42.2	54.2	42.2	38.0
	SVM (RBF)	38.3	44.8	38.3	37.2	76.2	76.8	76.2	76.4
VGG16	Bayes	40.7	41.1	40.7	39.7	51.0	51.2	51.0	50.0
	MLP	57.2	57.7	57.2	57.4	64.6	64.5	64.6	64.5
	KNN	40.8	44.0	40.8	40.3	53.8	55.6	53.8	53.7
	RF	53.7	53.6	53.7	52.5	66.6	67.1	66.6	66.4
	SVM (Poly)	27.0	31.7	27.0	19.4	45.4	55.6	45.4	41.9
	SVM (RBF)	61.0	61.2	61.0	61.0	67.6	68.3	67.6	67.8
VGG19	Bayes	40.5	40.1	40.5	38.6	50.1	50.6	50.1	49.2
	MLP	55.8	56.6	55.8	56.1	65.8	65.9	65.8	65.7
	KNN	38.7	41.9	38.7	38.6	51.0	54.2	51.0	51.2
	RF	54.2	53.8	54.2	53.0	62.6	63.1	62.6	62.6
	SVM (Poly)	31.6	34.5	31.6	23.7	38.8	39.9	38.8	32.8
	SVM (RBF)	56.3	56.5	56.3	56.3	66.8	67.8	66.8	67.1

5 Conclusion

In this work, different approaches to fingerprint classification were evaluated according to Henry's classification system. Among the different scenarios, techniques for extracting traditional features and based on deep features were evaluated. Notably, the conventional extractors' methods did not present satisfactory results, with an accuracy below 30%, on average. The methodology following feature extraction via CNN showed up to 78% accuracy, being superior to the traditional technique.

These results, however, do not exceed the scenario of using CNN as an image classifier. In the latter case, this work indicates that applying prior filtration with a Gabor filter combined with the thinning technique can contribute to results in some cases.

In the future, we plan to propose a new CNN architecture to solve fingerprint classification and recognition problems. Besides that, we plan to provide a complete evaluation with a time comparison between approaches. Other motivations arise from improving fingerprint recognition systems since acquisition is not always practical.

References

1. Abbood, A.A., Sulong, G., Razzaq, A.A.A., Peters, S.U.: Segmentation and enhancement of fingerprint images based on automatic threshold calculations. In: Saeed, F., Gazem, N., Patnaik, S., Saed Balaid, A.S., Mohammed, F. (eds.) IRICT 2017. LNDECT, vol. 5, pp. 400–411. Springer, Cham (2018). https://doi.org/10.1007/978-3-319-59427-9_43
2. Ahmad, F., Mohamad, D.: A review on fingerprint classification techniques. In: 2009 International Conference on Computer Technology and Development, vol. 2, pp. 411–415 (2009)
3. Almajmaie, L., Ucan, O.N., Bayat, O.: Fingerprint recognition system based on modified multi-connect architecture (MMCA). Cogn. Syst. Res. **58**, 107–113 (2019)
4. Cao, K., Pang, L., Liang, J., Tian, J.: Fingerprint classification by a hierarchical classifier. Pattern Recogn. **46**(12), 3186–3197 (2013)
5. Chhabra, M., Shukla, M.K., Ravulakollu, K.K.: Bagging- and boosting-based latent fingerprint image classification and segmentation. In: Gupta, D., Khanna, A., Bhattacharyya, S., Hassanien, A.E., Anand, S., Jaiswal, A. (eds.) International Conference on Innovative Computing and Communications. AISC, vol. 1166, pp. 189–201. Springer, Singapore (2021). https://doi.org/10.1007/978-981-15-5148-2_17
6. Ding, S., Shi, S., Jia, W.: Research on fingerprint classification based on twin support vector machine. IET Image Process. **14**(2), 231–235 (2020)
7. Ezzati Khatab, Z., Hajihoseini Gazestani, A., Ghorashi, S.A., Ghavami, M.: A fingerprint technique for indoor localization using autoencoder based semi-supervised deep extreme learning machine. Signal Process. **181**, 107915 (2021)
8. Gabor, D.: Theory of communication. part 1: the analysis of information. J. Inst. Electr. Eng. - Part III: Radio Commun. Eng. **93**(12), 429–441 (1946)
9. Gonzalez, R.C., Woods, R.E.: Digital Image Processing, 3rd Edn. Prentice-Hall, Inc., USA (2006)

10. Haghighat, M., Zonouz, S., Abdel-Mottaleb, M.: Identification using encrypted biometrics. In: Wilson, R., Hancock, E., Bors, A., Smith, W. (eds.) CAIP 2013. LNCS, vol. 8048, pp. 440–448. Springer, Heidelberg (2013). https://doi.org/10.1007/978-3-642-40246-3_55
11. He, K., Zhang, X., Ren, S., Sun, J.: Deep residual learning for image recognition. In: Proceedings of the IEEE Conference on Computer Vision and Pattern Recognition, pp. 770–778 (2016)
12. Henry, E.R.: Classification and Uses of Fingerprints. Wyman and Sons Limited (1906)
13. Jain, A., Prabhakar, S., Hong, L.: A multichannel approach to fingerprint classification. IEEE Trans. Pattern Anal. Mach. Intell. **21**(4), 348–359 (1999)
14. Karu, K., Jain, A.K.: Fingerprint classification. Pattern Recogn. **29**(3), 389–404 (1996)
15. Klimczak, L., von Eschenbach, C.E., Thompson, P., Buters, J., Mueller, G.: Applying deep learning to pollen identification using metabolic fingerprints. J. Allergy Clin. Immunol. **147**(2, Supplement), AB83 (2021). programs and Abstracts of Papers to be Presented During Virtual Scientific Sessions: 2021 AAAAI Virtual Annual Meeting
16. Krizhevsky, A.: Convolutional deep belief networks on cifar-10 (May 2012)
17. Krizhevsky, A., Sutskever, I., Hinton, G.E.: Imagenet classification with deep convolutional neural networks. Commun. ACM **60**(6), 84–90 (2017)
18. Lam, L., Lee, S., Suen, C.Y.: Thinning methodologies - a comprehensive survey. IEEE Trans. Pattern Anal. Mach. Intell. **14**, 869–885 (1992)
19. Le, N.T., Wang, J.W., Le, D.H., Wang, C.C., Nguyen, T.N.: Fingerprint enhancement based on tensor of wavelet subbands for classification. IEEE Access **8**, 6602–6615 (2020)
20. Li, J., Yau, W.Y., Wang, H.: Combining singular points and orientation image information for fingerprint classification. Pattern Recogn. **41**(1), 353–366 (2008)
21. Lin Hong, Yifei Wan, Jain, A.: Fingerprint image enhancement: algorithm and performance evaluation. IEEE Trans. Pattern Anal. Mach. Intell. **20**(8), 777–789 (1998)
22. Liu, M.: Fingerprint classification based on adaboost learning from singularity features. Pattern Recogn. **43**(3), 1062–1070 (2010)
23. Liu, S., Deng, W.: Very deep convolutional neural network based image classification using small training sample size. In: 2015 3rd IAPR Asian Conference on Pattern Recognition (ACPR), pp. 730–734 (2015)
24. Medeiros, A.G., et al.: A novel approach for automatic enhancement of fingerprint images via deep transfer learning. In: 2020 International Joint Conference on Neural Networks (IJCNN), pp. 1–8. IEEE (2020)
25. Miranda, N.D., Novamizanti, L., Rizal, S.: Convolutional neural network pada klasifikasi sidik jari menggunakan resnet-50. Jurnal Teknik Informatika (Jutif) **1**(2), 61–68 (2020)
26. Redmon, J.: Darknet: open source neural networks in c. http://pjreddie.com/darknet/ (2013–2016)
27. Rim, B., Kim, J., Hong, M.: Fingerprint classification using deep learning approach. Multimedia Tools Appl. 1–17 (2020). https://doi.org/10.1007/s11042-020-09314-6
28. Schuch, P., Schulz, S., Busch, C.: Survey on the impact of fingerprint image enhancement. IET Biometrics **7**(2), 102–115 (2018)
29. Serafim, P.B.S., et al.: A method based on convolutional neural networks for fingerprint segmentation. In: 2019 International Joint Conference on Neural Networks (IJCNN), pp. 1–8 (2019)

30. Sharma, R.P., Dey, S.: Two-stage quality adaptive fingerprint image enhancement using fuzzy c-means clustering based fingerprint quality analysis. Image Vis. Comput. **83–84**, 1–16 (2019)
31. Vasavi, S., Priyadarshini, N.K., Vardhan, K.H.: Invariant feature based darknet architecture for moving object classification. IEEE Sens. J. **21**(10), 11417–11426 (2020)
32. Wang, J.W.: Classification of fingerprint based on traced orientation flow. In: 2010 IEEE International Symposium on Industrial Electronics. IEEE (July 2010)
33. Wang, J.W., Le, N.T., Wang, C.C., Lee, J.S.: Enhanced ridge structure for improving fingerprint image quality based on a wavelet domain. IEEE Sig. Process. Lett. **22**(4), 390–394 (2015)
34. Watson, C.I., Wilson, C.L.: Nist special database 4. Fingerprint Database Natl. Inst. Stan. Technol. **17**(77), 5 (1992)
35. Win, K.N., Li, K., Chen, J., Viger, P.F., Li, K.: Fingerprint classification and identification algorithms for criminal investigation: a survey. Future Gener. Comput. Syst. **110**, 758–771 (2020)

Segmentation of Significant Regions in Retinal Images: Perspective of U-Net Network Through a Comparative Approach

Matej Pirhala(✉), Jozef Goga, Veronika Kurilova, Jarmila Pavlovicova, and Slavomir Kajan

Faculty of Electrical Engineering and Information Technology STU in Bratislava, Ilkovicova 3, 81219 Bratislava, Slovakia
{xpirhala,jozef.goga,veronika.kurilova,jarmila.pavlovicova, slavomir.kajan}@stuba.sk

Abstract. The main aim of this article is to compare U-Net models with different methods of medical image segmentation in ophthalmology when searching for the optic disc and vessels on the retina. We used freely available databases to analyze the results. DRIVE, STARE and MESSIDOR were used for blood vessel segmentation and DRION database was used for optic disc segmentation. In vascular segmentation, we compared the method based on morphological operations with U-Net convolutional network. The best results were obtained by neural network, where we achieved accuracy of 96.7% on DRIVE dataset. Methods based on morphological operations, Attentional U-Net network and Fast Radial Symmetry Transform were used for optic disc localization, where the U-net network achieved the best sensitivity of 91.6%. We show that in our tests U-Net architectures achieve better results in vessel and optic disc segmentation even compared to standard methods, that do not depend on the amount of available data.

Keywords: Attention U-Net · FRST · Morphological operations · Optic disc · Vessel segmentation · Ophthalmology

1 Introduction

Over time, several algorithms and methodologies have been developed for automatic identification, localization, and extraction of anatomical structures of the human retina. We can divide these approaches into two basic groups, rule-driven techniques, composed of combination of deterministic algorithms and machine learning approaches. There are many methods based on deterministic algorithms, such as the kernel-based method that compares pixel intensity variations in combination with the retinal cross-sectional profile [1]. Different filtering algorithms were developed to extract the vascular structure. Wu et al. [2] used a combination of Hessian and matched filters, which were applied to improve the contrast

G. Rozinaj and R. Vargic (Eds.): IWSSIP 2021, CCIS 1527, pp. 29–40, 2022.
https://doi.org/10.1007/978-3-030-96878-6_3

between the vascular structure and other parts of the retina. These methods are mostly based on pixel intensity monitoring. Mathematical morphology concerns more on the structure and shape of objects in the image. Jiang et al. [3] presented the work of vascular structure extraction using global thresholding based on morphological operations. This method achieved an average accuracy of 95.88% on merged (STARE+DRIVE) dataset and 95.27% for cross-dataset evaluation when trained on DRIVE and tested on STARE database. Segmentation of the optic disc is mostly based on pixel contrast analysis, as it is usually the region with the highest pixel intensity [4]. Important feature of the optic disc is also its shape and structure, which is approximately circular. Popular approaches for circle detection are FRST [5] and the Hough transform [6,7]. Morphological operations are also frequently used methods for an optic disc localization [8].

In recent years, machine learning methods have come to the fore. These are often convolutional networks that can solve the task effectively if they have enough training data. However, sufficient data is often a problem, especially in medicine. Ronneberger et al. [9] recently designed the U-Net architecture, which enables efficient deployment in biomedical applications for image segmentation through the ability to train effectively using very few images. However, these architectures are usually compared to deep learning approaches, i.e. standard convolutional networks. In this work, we therefore focus on the evaluation of U-Net models in comparison with standard approaches, which are not dependent on the large number of available training data.

2 Methods

2.1 Evaluation Techniques

The ability of the segmentation algorithm to correctly distinguish the structure of retinal vessels and optic disc can be determined using different metrics. The most often used metrics in medical applications are TPR (True Positive Rate), FPR (False Positive Rate), average specificity, accuracy, sensitivity and F1 score, which is also known as the Sørensen - Dice coefficient (DSC) [10]. The vessel segmentation was evaluated on the pixel level and for optic disc we measured the percentage of intersection of ground truth region with the resulting segment. In our experiments we used free available databases with marked vessels:

DRIVE: this database contains 40 color images of healthy retina with size 768 × 584 pixels. Each image contains a black and white mask of vessels [11].

STARE: this database contains 20 color 700 × 604 pixel image of the retina with black and white masks of vascular structure. This database contains two sets of masks for blood vessels, one created by ophthalmologist A. Hoover and the second by V. Kouznetsova. We compared our algorithms with masks created by A. Hoover. The images in this database often contain some degree of pathology, such as exudates or lesions [12].

MESSIDOR: this database was created thanks to funding from the French Ministry for research and defense [13]. The database contains 1200 images in TIF format in 3 different resolutions: 1440 × 960, 2240 × 1488 and 2304 × 1536 pixels. As this database contains for each image only the degree of disease and does not contain vascular masks, thanks to ophthalmologist V. Kurilova, were created for the three images with a resolution of 2240 × 1488 pixels a black and white masks of the vascular structure of the retina [14].

The optic disc segmentation methods were tested on DRION database [15] with 109 images with resolution 600 × 400 pixels, where optic disc segments are assigned as binary image masks for each image.

Each of these databases were divided according to 3 fold-cross validation approach including one held-out set. Images in every database were randomly shuffled and divided into 4 sets. Three sets were assigned for the training and one static set for the held-out test. DRIVE database was divided into 10 images intended for training, 10 for validation and 20 for testing. The training set of STARE database contained 5 images. Validation set contained also 5 images and test set was comprised of 10 images. Since MESSIDOR database contained only three images with the evaluated vascular masks, we have assigned all the images to the test set. We used the same data sets for training and validation as for the STARE database.

The 90 images were randomly divided into 3 groups of 30 images for training and validation. The fourth group with 19 images was used for testing.

2.2 Vessel Segmentation

We tested two methods of vessel segmentation. We compared method based on mathematical morphology that selects the vascular features by suitable structural elements [14] with the deep learning U-Net model, based on compression and reconstruction of image data.

Method Based on Morphological Operations. For our comparison, we used the method described in [14]. The green component of the color image was selected and normalized according to the intensity of pixels. The segmentation algorithm run on green channel in two independent threads with similar operations, but different structural elements. Various structural elements define different vessel characteristics. Smaller elements define thinner vessels, they could be distorted by larger elements. The achieved results of single threads were subsequently combined. For binary segmentation, the Otsu's thresholding method [16] was used. Finally, the rotated linear structural elements with rotation step 4° in 60 directions were applied.

Method Based on U-Net Convolutional Network. The principle of U-Net model is image processing in two stages. In the first stage, the image data are reduced progressively to get the characteristic image features and in the second stage, subsequent image reconstruction is applied to segment the desired structure. In the first stage, we used a stochastic loss function proposed by Khanal et al. [17] to obtain an overall probability map for each pixel of the image. The goal of this map is not to segment the image directly, but to identify the probability of a particular pixel being frequented by a vessel. Image segmentation does not take place at this stage due to the downsampling used by the U-Net network. The original vessels are blurred, which makes it difficult to correctly mark thin vessels. We then identified those pixels that are most likely misclassified, such as narrow vessels and pixels at the edges of thick vessels. Subsequently, we applied a second convolutional network to these areas, the so-called Mini U-Net [17]. This Mini U-net uses a middle layer of U-Net architecture. Inputs to this network are two channels, a complete probability map (the output of the first network) and a version of this map that contains ambiguous pixels.

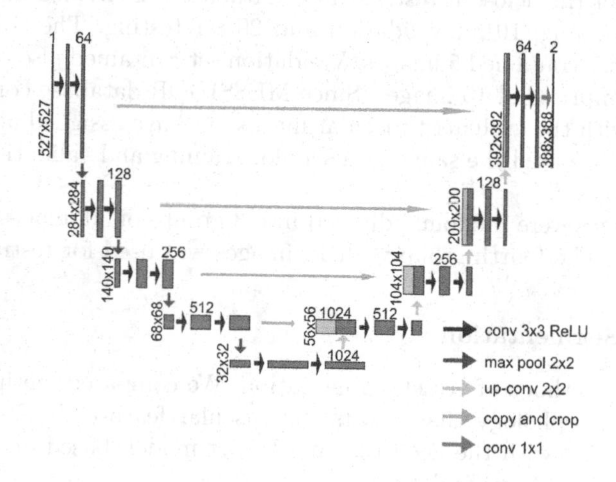

Fig. 1. Model of used U-Net neural network.

2.3 Optic Disc Segmentation

Optic disc is the region in retinal fundus image, where the optic nerve ends. It is usually the lightest approximately circular object.

Optic Disc Segmentation Based on Attention U-Net Convolution Network. We based this segmentation on research described in [18]. In this method, a U-Net network with attention was used. Proposed attention gates are included into the standard U-Net architecture (Fig. 1). Information extracted from the coarse scale image is used in gating to eliminate irrelevant and noisy responses via bridging connections. This is performed directly before the pooling layers to merge the relevant activations. In addition, attention gates filter neural activations during forward, as well as backward pass. Gradients originating from the background areas are weighted during backpropagation. This allows us to update the parameters of model in shallower layers mostly on spatial areas that are relevant for given task [18].

Optic Disc Segmentation Based on Morphological Operations. This method was inspired by research described in [19]. In the image pre-processing stage, the combination of channels of different colour image models was used. First, a 3×3 median filter was applied to the image. The saturation and value channels from image in HSV colour space were multiplied pixel by pixel by the red channel of the RGB image colour space. Subsequently, morphological closing was applied to resulting grayscale image. After this operation, contrast was enhanced by contrast limited adaptive histogram equalization (CLAHE) [20]. After equalization, global threshold was found. We were increasing treshold value until the number of white pixels in the thresholded image was less than 7500. This value was determined experimentally and achieved the best results for the given database. The thresholded image was processed by morphological filtering to remove white objects smaller than 2000 pixels. Also the black spots smaller than 3000 pixels were filled. Finally, the boundaries of the optic disc were determined by the Canny edge detector [21].

Method Based on Fast Radial Symmetry Transform - FRST. FRST is a transform based on radius symmetry, first introduced in [5]. This transform is calculated for one or more radii $n \in N$, where N is the set of radii of radially symmetric objects to be detected. For each radius n, two images are created, namely an oriented image O_n and a magnitude image M_n of the projected image. These images are generated by examining the gradient g, in each pixel p. In this method, we used the research from [22], where we replicated the process of segmentation of the optic disc accurately. In addition to the original method, we used the parameter γ, where γ is an approximation of the highest gradient at the boundaries of the optic disc. This means that the gradient value greater than γ is not considered to be a part of the optic disc. We then applied this method to the image, setting the γ parameter to 12, β in this case equals 0 and α is 0.01. We ran the algorithm several times to search for a radius in the range of 10 to 90 pixels with step 10. We obtained the center of the optic disc by finding the global maximum from the complete transformation S. We then determined the optic disc mask as a circle centered at a global maximum and a radius of 50 pixels.

3 Experiments and Results

3.1 Vessel Segmentation

Vessel Segmentation Based on Morphological Operations. The experiments were performed on a test samples from databases DRIVE, STARE and

MESSIDOR, on which we also determined the efficiency of neural network segmentation.

The Table 1 contains a summary comparison of this method on all databases. This method was optimized for high resolution images and in accordance with this, it achieved the best results on MESSIDOR database.

Table 1. Evaluation metric for vessel segmentation by morphological operations on the test datasets.

Database	DRIVE	STARE	MESSIDOR
F1	0.585	0.588	0.750
Sensitivity	0.431	0.458	0.724
Specifity	0.996	0.992	0.992
Positive predictive value	0.921	0.839	0.778
Accuracy	0.945	0.948	0.981

In mathematical morphology, the size of either the structural element or the segmented parts of the image were particularly important. It is therefore obvious that even vessels that would have a width of a few pixels in the STARE or DRIVE database, in MESSIDOR images they have width of many more pixels, and therefore they were not filtered out during morphological operations.

In particular, thick vessels were segmented on the STARE and DRIVE databa-ses, and morphological operations filtered out thin veins and capillaries, which led to a deterioration in the results.

Vessel Segmentation Based on U-Net Convolution Network. Based on the principle of cross-validation, we prepared three sets from each database and the resulting value of evaluation metrics were averaged from the outputs of each dataset. For training on the DRIVE database, 1120 images were created. They were evaluated during training with a batch size of 2 images. The input images had dimensions of 572 × 572 pixels, but only the 388 × 388 area was evaluated. For the STARE database, we created 1080 input training images. Images were created to enlarge the training set based on sliding window method with overlap and step of 20 pixels.

Since MESSIDOR contains only three images with the evaluated vascular masks, we determined all these images as test images. The masks of this database were created by ophthalmologist. We used the same datasets for training as for the STARE database.

In total, the network was trained by standard backpropagation algorithm minimizing categorical cross-entropy loss using Adam optimizer for the fixed number of 5 epochs and default hyperparameters for each database. Due to limitations, we used mini-batch size 2 with offline augmentation in an effort to improve the visibility of the searched objects in the retinal images. First, the input images were resized as needed and per channel normalized over the dataset by subtracting the mean and dividing by standard deviation. Subsequently, we normalized the images to the range $< 0, 1 >$ by maximum value division (Figs. 2, 3, 4, 5, 6, 7, 8, 9 and Table 2).

Table 2. Evaluation metric for vessel segmentation by U-Net convolutional network on test datasets.

Database	DRIVE	STARE	MESSIDOR
F1	0.817 ± 0.003	0.761 ± 0.024	0.474 ± 0.106
Sensitivity	0.828 ± 0.022	0.673 ± 0.043	0.374 ± 0.108
Specificity	0.981 ± 0.003	0.993 ± 0.002	0.994 ± 0.0009
Positive predictive value	0.810 ± 0.020	0.899 ± 0.021	0.756 ± 0.012
Accuracy	0.967 ± 0.0009	0.967 ± 0.002	0.971 ± 0.003

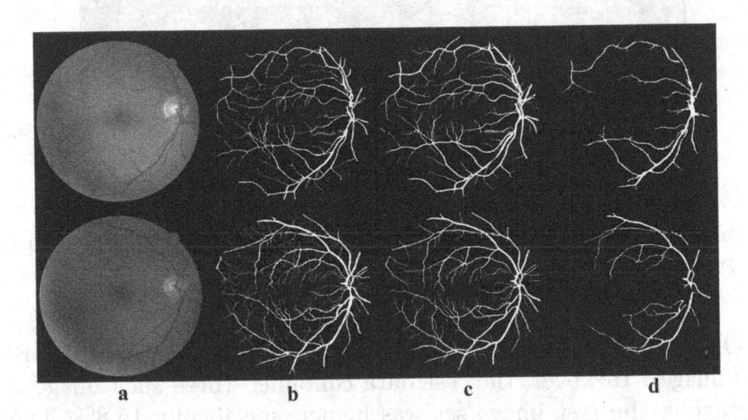

Fig. 2. [From left to right] a) Original images of DRIVE database b) Ideal masks c) Segmentiation with U-net network d) Segmentation with morfological operation method.

3.2 Optic Disc Segmentation

Optic Disc Segmentation Based on Attention U-Net Convolution Network. We applied three fold cross-validation to correctly evaluate the neural network performance on all used datasets. Again, 800 network input images were generated from each training data with sliding window method as in the Sect. 3.1. For each image used, we ensured there existed the black-and-white optic disc target mask, as the ground truth image for the prediction verification. The training was performed with fixed number of 10 epochs using the same hyperparameters as in previous experiments. In Table 3 we present the results of individual metrics of the cross validation. The value of the standard deviation is negligibly small in all cases, which indicates stability of this approach.

Table 3. Evaluation results for optic disc segmentation.

Method	F1	Sensitivity	Specificity	Positive predictive value	Accuracy
Attentional U-Net	0.917 ± 0.02	0.916 ±0.010	0.998 ± 0.0009	0.918 ± 0.028	0.995 ± 0.001
Morphological operations	0.773	0.738	0.996	0.817	0.988
FRST	0.486	0.503	0.982	0.472	0.967

Optic Disc Segmentation Based on Morphological Operations. This method has obtained satisfactory results, which can be seen in Table 3.

The resulting images can then be divided into three groups:

1. **Correctly Specified Images:** These images had high metric values and were segmented with sufficient quality. Most of such images were from DRION database, specifically 91.8%. However, the share of these images in the test data dropped to 73.7%.

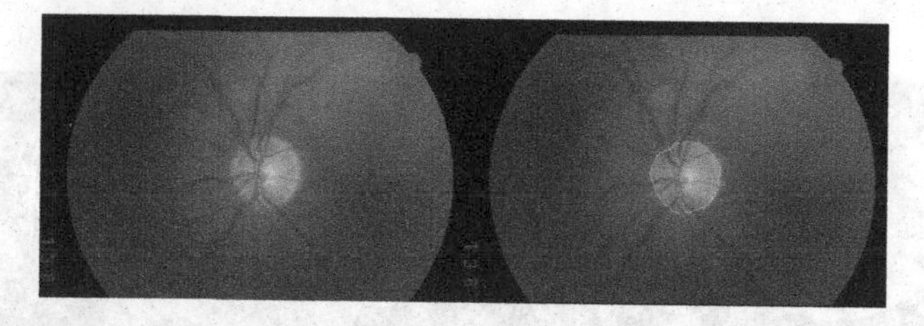

Fig. 3. Example of DRION image (left) and algorithm output (right) with correctly specified optic disc.

2. **Two Optic Disc Locations Specified:** This option accounted for 5.5% in all 109 images. However, the test data contained three such images, i.e. the percentage in the test image set was higher, specifically 15.8%. The reason for these results was probably the phenomenon that the surroundings had a similar intensity as the optic disc. To minimize these conditions, a ring search method such as FRST (Fast Radial Symmetry Transform) or Hough Transform could be used to decide which feature is more likely to be an optic disc.

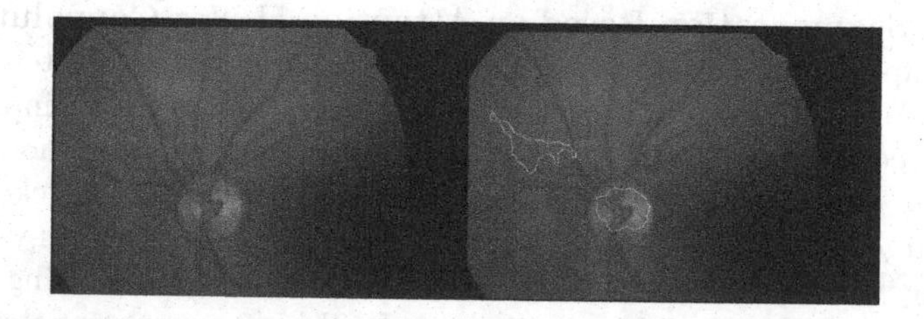

Fig. 4. Example of DRION image (left) and algorithm output (right) with two optic disc segmentation.

3. **Incorrectly Determined Images:** There were four incorrectly determined images in the entire DRION dataset, two of which were in the test data, which is about 10.5% in the test data and 3.7% in the entire database. This was due to the fact that the green channel contained higher ambient pixel intensities than the optic disc itself. It is also caused by the value of 7500, if this value was higher, then these images were qualified as correctly determined.

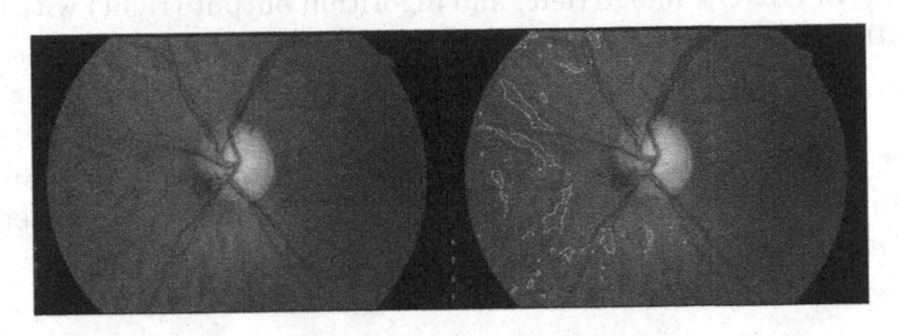

Fig. 5. Example of DRION image (left) and algorithm output (right) with incorrect optic disc segmentation.

Optic Disc Segmentation Based on Fast Radial Symmetry Transform. We tested this method on 109 DRION database images. To compare this method with neural networks, we used the same datasets as in 3 cross-validations for neural networks. However, since this is a deterministic algorithm we only compare the test dataset of images. The overall accuracy on the test data set by the FRST method can be seen also in Table 3.

1. **Correct Localization:** these images showed minimal deviation. Their metrics were relatively high. There were 31.2% of such images, of all 109 images.

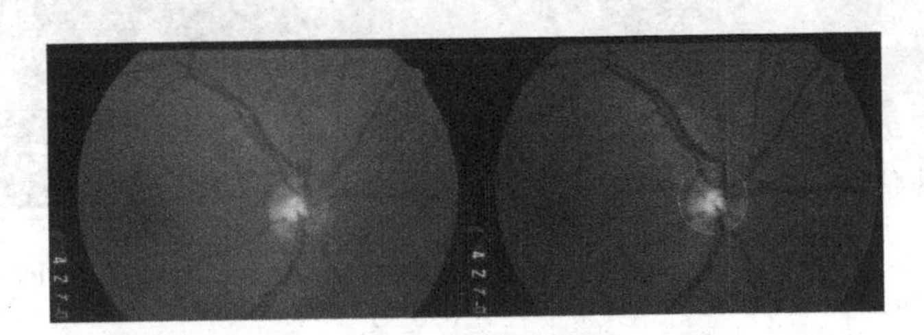

Fig. 6. Example of DRION image (left) and algorithm output (right) with correct optic disc segmentation.

2. **Correct Localization with Shifted Optic Disc:** The result of these images showed significant deviations in the centers of the optic discs, and the optic disc itself was shifted by a certain distance. However, part of the localized disc was in the disc mask of the image.

Fig. 7. Example of DRION image (left) and algorithm output (right) with shifted optic disc segmentation.

3. **Incorrect Localization:** these images accounted for 11.9% of the entire image database. These images showed zero, or a value close to zero for sensitivity, F1 coefficient and positive predictive value.

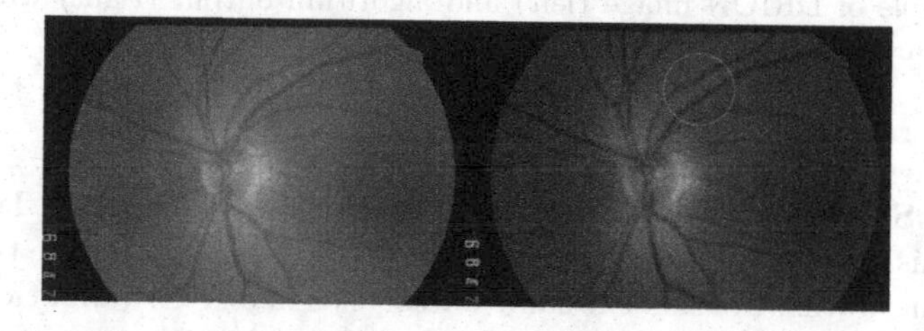

Fig. 8. Example of DRION image (left) and algorithm output (right) with incorrect optic disc segmentation.

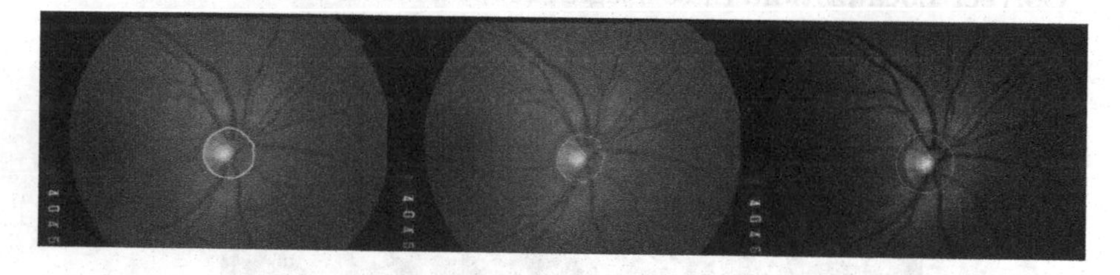

Fig. 9. Localization of optic disc with U-Net network (left), morphological operation method (in the middle) and FRST method (right).

4 Conclusion

A better method in the case of testing and training on DRIVE database is the U-Net neural network, as it was able to more accurately identify capillaries. In the case of evaluation on the STARE database, the neural network was again better than standard methods. Again, morphological operations failed to segment fine vessels. This could be due to the size of the structural elements used in the method. In the case of the MESSIDOR database, we can see that morphological operations achieved the best segmentation. There are few reasons for this result. First, the algorithm was optimized for high resolution MESSIDOR dataset and so it achieved worse results on DRIVE and STARE datasets. Second, there are only 3 images with marked vessels, that would not be enough data to train the U-Net network. Because of that, U-Net was trained only on lower resolution images from DRIVE and STARE. These two segmentations were comparable to the naked eye. However, we assume that if the U-Net network could be trained on a larger amount of data and hyperparameters would be tuned, results would be further improved.

As we can see in the results of individual methods, the Attention U-Net network achieved best results compared to standard methods that do not depend on the amount of data available. In addition, U-Net network was able to accurately determine the correct shape of the optic disc, with no completely poorly segmented image.

Acknowledgement. This paper and the research behind has been supported by Grant 1/0611/22 of the Slovak Scientific Grant Agency VEGA.

References

1. Chaudhuri, S., Chatterjee, S., Katz, N., Nelson, M., Goldbaum, M.: Detection of blood vessels in retinal images using two-dimensional matched filters. IEEE Trans. Med. Imaging **8**(3), 263–269 (1989)
2. Wu, C.-H., Agam, G., Stanchev, P.: A general framework for vessel segmentation in retinal images. In: 2007 International Symposium on Computational Intelligence in Robotics and Automation, pp. 37–42, IEEE, Rupnagar, India (2007). https://doi.org/10.1109/CIRA.2007.382924
3. Jiang, Z., Yepez, J., An, S., Ko, S.: Fast, accurate and robust retinal vessel segmentation system. Biocybernetics Biomed. Eng. **37**(3), 412–421 (2017)
4. Walter, T., Klein, J.-C.: Segmentation of color fundus images of the human retina: detection of the optic disc and the vascular tree using morphological techniques. In: Crespo, J., Maojo, V., Martin, F. (eds.) ISMDA 2001. LNCS, vol. 2199, pp. 282–287. Springer, Heidelberg (2001). https://doi.org/10.1007/3-540-45497-7_43
5. Loy, G., Zelinsky, A.: A fast radial symmetry transform for detecting points of interest. In: Heyden, A., Sparr, G., Nielsen, M., Johansen, P. (eds.) ECCV 2002. LNCS, vol. 2350, pp. 358–368. Springer, Heidelberg (2002). https://doi.org/10.1007/3-540-47969-4_24
6. Tjandrasa, H., Wijayanti, A., Suciati, N.: Optic nerve head segmentation using hough transform and active contours. Telkomnika **10**(3), 531–536 (2012). https://doi.org/10.12928/telkomnika.v10i3.833
7. Lesay, B., Pavlovicova, J., Oravec, M., Kurilova, V.: Optic disc localization in fundus image. In: IWSSIP 2016: 23th International Conference on Systems. Signals and Image Processing, pp. 23–25. Bratislava, Slovakia (2016)
8. Ashwitha, K., Srikanth, R.: Morphological background detection and enhancement of images. Int. J. Innovative Technol. Res. **4**(6), 4466–4471 (2016)

9. Ronneberger, O., Fischer, P., Brox, T.: U-net: convolutional networks for biomedical image segmentation. In: Navab, N., Hornegger, J., Wells, W.M., Frangi, A.F. (eds.) MICCAI 2015. LNCS, vol. 9351, pp. 234–241. Springer, Cham (2015). https://doi.org/10.1007/978-3-319-24574-4_28
10. Chicco, D., Jurman, G.: The advantages of the Matthews correlation coefficient (MCC) over F1 score and accuracy in binary classification evaluation. BMC Genomics **21**(1), 1–13 (2020). https://doi.org/10.1186/s12864-019-6413-7
11. Staal, J., Abramoff, M.D., Niemeijer, M., Viergever, M.A., van Ginneken, B.: Ridge-based vessel segmentation in color images of the retina. IEEE Trans. Med. Imaging **23**(4), 501–509 (2004)
12. Hoover, A.D., Kouznetsova, V., Goldbaum, M.: Locating blood vessels in retinal images by piecewise threshold probing of a matched filter response. IEEE Trans. Med. Imaging **19**(3), 203–210 (2000)
13. Decencière, E., et al.: Feedback on a publicly dsitributed image database: the MESSIDOR database. Image Anal. Stereology **33**(3), 231 (2014)
14. Kurilova, V., Pavlovicova, J., Oravec, M., Rakar, R., Marcek, I.: Retinal blood vessels extraction using morphological operations. In: 2015 International Conference on Systems, Signals and Image Processing (IWSSIP), pp. 265–268, September 2015. https://doi.org/10.1109/IWSSIP.2015.7314227
15. Carmona, E.J., Rincón, M., García-Feijoó, J., Martínez-de-la-Casa, J.M.: Identification of the optic nerve head with genetic algorithms. Artif. Intell. Med. **43**(3), 243–259 (2008)
16. Otsu, N.: A threshold selection method from gray-level histograms. IEEE Trans. Syst. Man Cybern. **9**(1), 62–66 (1979)
17. Khanal, A., Estrada, R.: Dynamic deep networks for retinal vessel segmentation. arXiv preprint arXiv:1903.07803 (2019)
18. Oktay, O., et al.: Attention u-net: learning where to look for the pancreas. arXiv preprint arXiv:1804.03999 (2018)
19. Pal, S., Chatterjee, S.: Mathematical morphology aided optic disk segmentation from retinal images. In: 2017 3rd International Conference on Condition Assessment Techniques in Electrical Systems (CATCON), pp. 380–385 (2017). https://doi.org/10.1109/CATCON.2017.8280249
20. Reza, A.M.: Realization of the contrast limited adaptive histogram equalization (CLAHE) for real-time image enhancement. J. VLSI Sig. Process. Syst. Sig. Image Video Technol. **38**(1), 35–44 (2004)
21. Canny, J.: A computational approach to edge detection. IEEE Trans. Pattern Anal. Mach. Intell. **6**, 679–698 (1986)
22. Budai, A., Aichert, A., Vymazal, B., Hornegger, J., Michelson, G.: Optic disk localization using fast radial symmetry transform. In: Proceedings of the 26th IEEE International Symposium on Computer-Based Medical Systems, pp. 59–64 (2013). https://doi.org/10.1109/CBMS.2013.6627765

Presenting a System to Aid on the Examination of Scintigraphy Bone Analysis Using DICOM Files

José M. Carneiro da Silva[1(✉)], Fernando Fernandes[2], Heron Botelho[2], Cláudio T. Mesquita[2], and Aura Conci[1]

[1] Universidade Federal Fluminense, Av. Milton Tavares de Souza, Niterói, Rio de Janeiro 24210346, Brazil
josemorista@id.uff.br

[2] Hospital Universitário Antônio Pedro, Av. Marquês de Paraná, Niterói, Rio de Janeiro 24033900, Brazil

Abstract. This work presents the use of computer vision and machine learning techniques to implement a semi-automated system to aid on understanding scintigraphy DICOM exams. In such a way, all aspects included from the image preprocessing to the identification and final classification of bone anomalies by the digital scintigraphy images are presented in details.

Keywords: Bone scintigraphy · Medical images processing · Bone metastasis

1 Introduction

Metastases can develop when cancer cells break away from the primary tumor and enter the system of carry fluids around the body (i.e. the bloodstream or lymphatic system) [1].

To perform the identification of bone metastasis from a primary cancer, is necessary a systematic inspection of scintigraphy images by trained and expert physician. However, in the majority of occasions, this process is performed only through visual analysis without quantitative assessment of the extent and severity of these metastases.

This aforementioned way of examination makes impractical to compare the progression of the patient over the time in a more objective way, since results based only on the deduction of a professional in a given inspection may not be faithfully reproducible at a future examination by himself and even less coincident with the evaluation of another professional.

The goal of this research is to verify the viability of using image analysis, computer vision and machine learning techniques to propose a system that is capable of assisting a qualified professional in bone metastases analysis.

G. Rozinaj and R. Vargic (Eds.): IWSSIP 2021, CCIS 1527, pp. 41–52, 2022.
https://doi.org/10.1007/978-3-030-96878-6_4

The work presented in this paper had the participation of resident students of the postgraduate course in nuclear medicine at Universidade Federal Fluminense - UFF. These colleagues were coauthors in the development and advisers in the stages related to the understanding and manipulation of medical and biological data, providing, with their high expertise, subsidies in the construction of the system.

To accomplish the proposed goal, the application must perform various previous stages linked to the process. The system stages are: (1) the management of the digital files of the exams stored in the Digital Imaging and Communications in Medicine (DICOM) format after their acquirement, (2) the pre processing of the original images that compose an exam, (3) the segmentation of these pre processed images, (4) the identification of the important elements in the segmented parts, (5) classification of these elements and (6) finally the quantification of bone anomalies that can be found using the mentioned digital image processing and machine learning techniques. All such stages will be detailed in next sections.

2 Bone Scintigraphy

Although there are three-dimensional imaging techniques combining information from scintigraphy with computed tomography like the Single Photon Emission Computed Tomography (SPECT) and Positron Emission Tomography (PET) [2], for bone metastases, scintigraphy is still one the most common imaging procedures in nuclear medicine, according to the European Association of Nuclear Medicine (EANM) [3].

Bone scintigraphy is a particularly important method for the clinical diagnosis of metastases [4]. When other examination methods are unable to provide a reliable diagnosis, scintigraphy becomes the most suitable means of making a final conclusion [4–6].

Scintigraphy allow the bone metastasis diagnosis by representing the physiological response to the active element used, making it possible to highlight the regions (in the whole skeleton) where this element is being more absorbed [6] (i.e. bone abnormalities areas are related to intensive activity of radionuclides). To perform this examination, a venous injection composed of a pharmaceutical agent called a radiotracer is applied to the patient. This application can occur up to 3 h before the scintigraphic image is collected (this is an exam that has a long duration).

As previously described, using a purely visual approach, examining each of the regions of intensive radio nuclide activity becomes a complex task even for specialized professionals. This complexity is due to the number of tasks that must be performed, since they must be able to visually identify, quantify and classify the areas of interest in the image from the exam. In addition to the inherent difficulty of the approach, other factors such as low contrast and noise (present in the image) can make this analysis more difficult and impair correct visual quantification and diagnosis. The idea of this work is to use the computer to assist in this task, initially helping the organization of data and then facilitating physicians to carry out the various stages of the analysis of a scintigraphic exam. Examples of these are the improvement of the image quality by contrast enhancement, inclusion of automatic segmentation mechanisms to identify regions of greater activity, the collection of representative attributes from each of these identified regions and the use of a computational intelligence algorithm for the classification process.

The exams used in this study were provided by the physicians and researchers partners from the nuclear medicine course of the Universidade Federal Fluminense (UFF). More specifically, the data consist of DICOM scintigraphic exams, reports and information about real patients of the university hospital: Hospital Universitário Antonio Pedro (HUAP).

The Image Processing techniques and the methods used for machine learning will be discussed right after a quick bibliographic review, which will be done in the next section. In the last section, the results achieved and the ideas for improvement and continuation of this research will be commented.

3 Related Works for Scintigraphy Aid

An interesting work carried out on automation of the classification and quantification of bone metastases was the EXINI [7] project published in 2006 and later BONENAVI [8]. They investigated artificial neural networks (ANN) application in the use of the Bone Scan Index (BSI). BSI is a parameter introduced with the proposal of serving as a clinical, quantitative and reproducible means to measure the evolution of bone metastases [1].

At 2017, another project from the Lund University studied the usage of convolutional neural networks(CNNs) to classify bone scan hotspots as metastatic/non-metastatic. The dataset for this task contained a total of 10427 hotspots, of which 3169 are positive (high risk of metastasis) and 7258 are negative (low risk of metastasis) [9] and was also provided by EXINI. The best performing ensemble of CNNs gave a area under the receiver operating characteristic curve score of 0.97 [9].

More recently, in 2020, another study investigates the application of a CNN to classify bone metastasis using whole body images of men initially diagnosed with prostate cancer. The proposed method employs different CNN-based architectures with data normalization, data augmentation and shuffling [1]. At the end of the study the algorithm scored with a recall of 98%, accuracy of 97% and precision of 95%.

Analyzing previous works on this field, it is noted that there are a number of possibilities for new approaches and implementation strategies to promote better medical evaluation of this type of examination.

4 An Overview of the Implemented System

The developed system offers an intuitive interface so that physicians and specialized professionals can perform analysis, segmentation and classification of scintigraphic exam files in DICOM format.

The application interface uses the WWW and was implemented with React.Js, HTML, CSS3 and TypeScript technologies. The system communicates through HTTP (Hypertext Transfer Protocol) requests in JSON (JavaScript Object Notation) format with a server maintained in the cloud. This server, implemented with Node.Js, TypeScript, Express, PostgreSQL and Redis is responsible for handling and processing all requests coming from the web application, performing requested operations related to image processing and artificial intelligence. The server also runs sub-processes (children) implemented in Python. Python libraries used were: Pydicom for reading and processing files in DICOM format; Skimage, Opencv and Scipy to perform operations related to the analysis and processing of digital images and SkLearn and Keras to classification and learning tasks.

The source code of this project is open, free and accessible at the following online repositories: github.com/josemorista/bm-server (server) and github.com/josemorista/bm-web (web application). The implemented web application is also available online for public access at bm-diag.org.

5 Data Entry: Patient Management and Exams Upload

To enter the application, the user's authentication is checked (Fig. 1). Previously, the user must create his account by providing basic data such as e-mail address, name, institution, job and password for access. The password is encrypted using the Bcrypt function. After the account creation, the user can access the system with his email address and password.

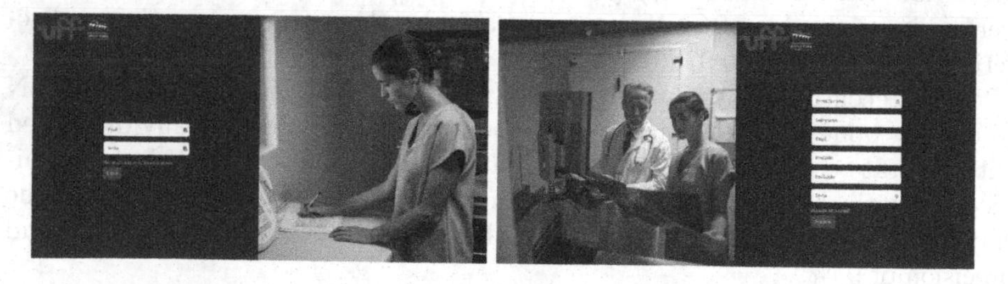

Fig. 1. View of the system homepage and registration.

Once authenticated in the platform, the user is redirected to a page where the already register patients can be found. To perform the insertion of a new patient in the system, it is necessary to include information such as the patient name, age, sex and complementary data such as previous cancer diagnoses, history of radiotherapy or chemotherapy treatments, previous injuries on the body and implants (please see Fig. 2).

By selecting one of the patients in the system, respective information and a link to access the list of available exams are presented (as shown in Fig. 2). In order to include new files to be stored for the same patient the user can click on the "Novo exame" button, that means "New exam" (the main language of this implementation version was Portuguese).

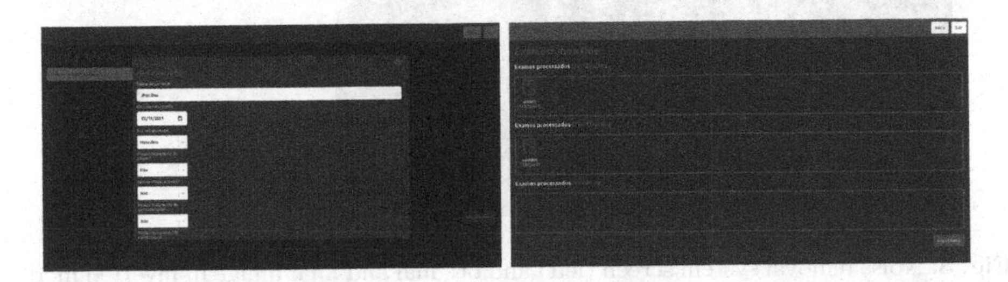

Fig. 2. View of the pages of new patient creation and exams.

After clicking this button, the user is redirected to the first stage of the assisted diagnostic flow, in which it is possible to provide additional information about the patient's position at the exam and upload the scintigraphic image file (.dcm). After submitting the form, the file is sent to the server and the corresponding exam created in the system database.

6 Preprocessing

When the file arrives in the server, it stores a copy of it in its own file system, so that it can carry out the various filtering, segmentation and learning operations that will be described in the following sections.

The first operation consists of a new linear transformation of the values of the absorption matrix extracted from the DICOM file in the range 0–4000 to floating point values in the range of 0 to 1. A simple filter performed in this step refers to the application of a maximum threshold (specified by the user) for the analysis of the file, thus, intensities above this threshold are replaced by the maximum value of 1. This allows the removal of a small group of pixels which intensities makes difficult to visualize the image file after its transformation.

After this conversion, the result image is presented to the user in order to carry on the desired image treatment operations. The available operations are the median, Gaussian and bilateral filters. If no option is selected, the default choice for this step is the usage of the median filter, since during the testing phase of the implementation it presented best results for the available images. For example, in Fig. 3 is possible to see the system screen referring to this operation and a comparison of a scintigraphy image in a inverted mode (that is image intensity

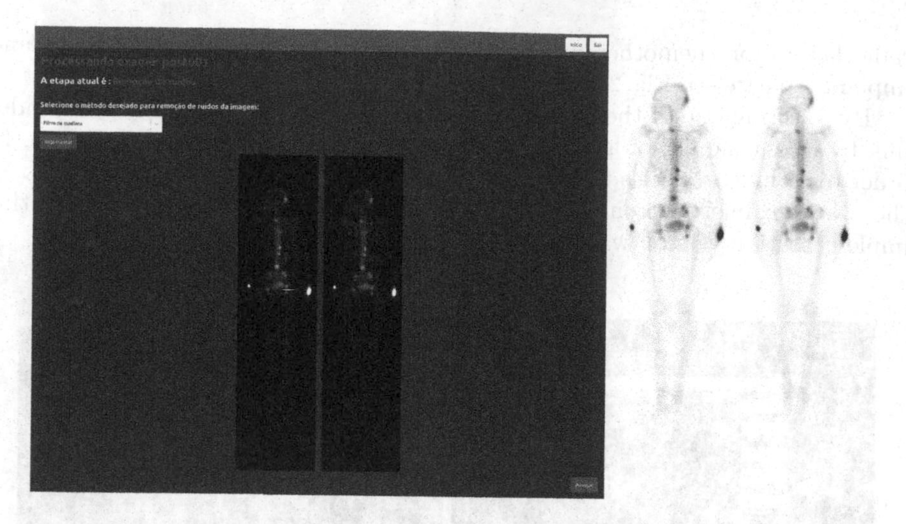

Fig. 3. Noise removal system screen (left) and original and final image in inverted mode (center and right).

level changed in it's gray level in order to promote better visualization of the differences) before and after executing the median filter.

After the image treatment is carried out, the segmentation phase begins. In this, a threshold related to the intensity of the gray tones of the pixels present in the image is defined, in order to promote better segmentation, leaving only the background and the detected anomalies. For this, by default the application recommends the use of a combination of segmentation through K-Means unsupervised learning followed by the Otsu thresholding algorithm for definition of the best intensity to binary limits. It is important to note that for the application of the K-Means technique, the user must provide the level of intensity ranges in which the wanted image will be presented, so the algorithm can perform the operation considering the desired number of clusters. Figure 4 shows the interface for this operation and the result of a processed and segmented image with the combination of K-Means and Otsu in inverted mode.

If the user is not satisfied with the result or does not wish to use the suggested algorithm, a manual thresholding can also be applied by the use of Random Walker algorithm. The process of applying the algorithms described in this phase can be repeated as wanted until the segmentation is considered satisfactory by the end user.

The next phase considers edge detection. At this point, the default method of the application is the Sobel filter, this standard choice is motivated by the good performance and computational cost of the method at some performed tests, however, other options are available such as Prewitt, Roberts and Scharr. Figure 5 illustrates the result obtained by this step in a inverted mode and the page in the system related to these operations.

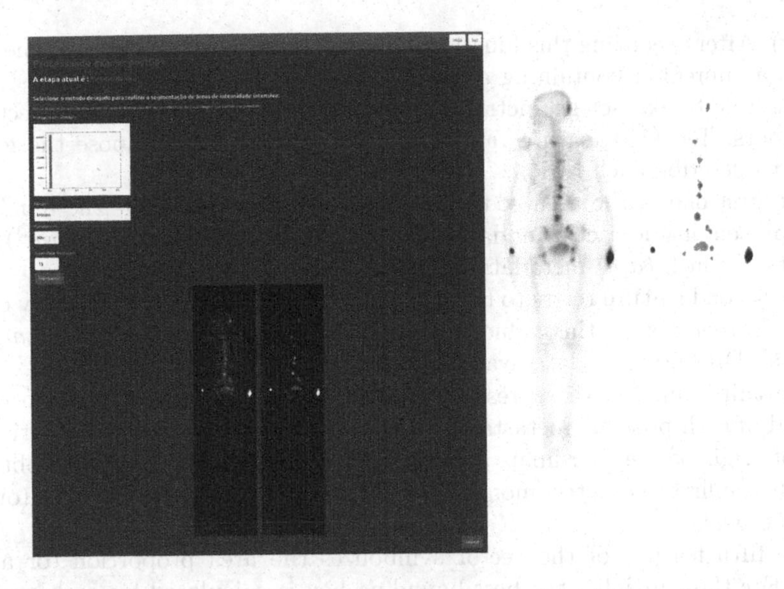

Fig. 4. Thresholding system screen (left) and original and final image in inverted mode (center and right).

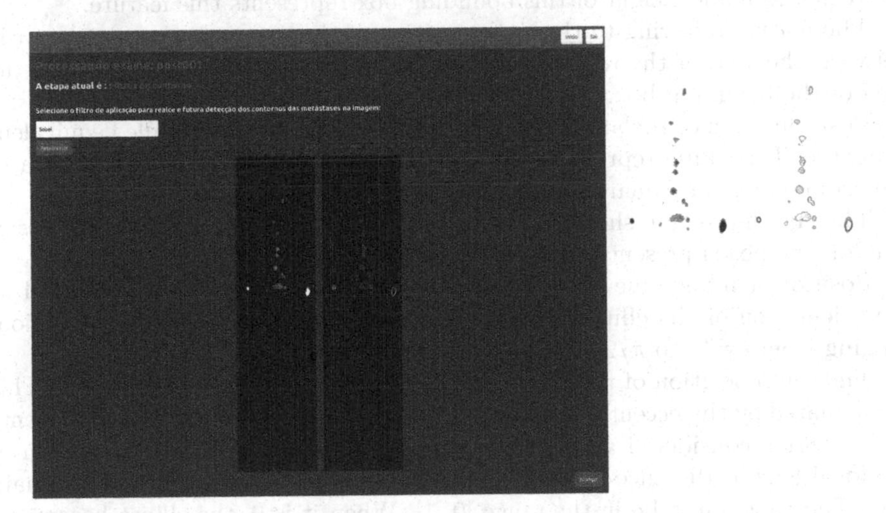

Fig. 5. Edge detection system screen (left) and original and final image in inverted mode (center and right).

7 Generation of Feature Vectors for Each Exam

Once an image with well-defined edges and regions is obtained, the process of generating the feature vectors that will be used in subsequent supervised learning tasks begins. For this purpose, was used the function *findContours* of the implementation of the algorithm described by Suzuki [10] (available in the OpenCV library). After executing this function, a list of entries is obtained, each one referring as a subregion containing a closed contour of the image, from this region it is possible to extract geometric information such as area, enclosing rectangle and others. Ten (10) features are computed and used to compose the feature vector to describe each region.

The first one is the area of the region. This is computed by the number of pixels present inside each boundary converted to square millimeters (mm^2) using the dots per inch (dpi) metadata available in the DICOM exam file.

The second feature refers to the perimeter of each region. Obtained by counting the border pixels, this value was later converted to millimeters (mm) also using the DICOM metadata available.

The third and fourth represent the horizontal and vertical position of the centroid of each possible metastasis. These values are calculated considering the point of origin of the coordinate system at the top left of the image and obtained through the first geometric moments of the region divided by the area (or zero order moment).

The fifth feature of the vector symbolizes the area proportion (or aspect ratio). For this, initially, the best bounding box is calculated for each one, and then its horizontal (width) and vertical (height) size are evaluated. The ratio between width and height of this bounding box represents this feature.

The feature referring to the sixth position of the vector represents the ratio between the area of the region (calculated in the first vector position) and the area of the bounding box.

Position seven of the attribute vector considers an attribute called equivalent diameter. This value represents the diameter of the circle of the same area of the grouped region (square root of the area divided by pi).

The eighth position shows the value of the average intensity of the shades of gray on the pixels present inside the grouped region.

Position number nine symbolizes the orientation of the region. Defined as the orientation of the ellipse that has the same second moments of the region, ranging from $-\pi/2$ to $\pi/2$ counterclockwise.

The tenth position of the vector refers to the eccentricity of the region. This is calculated by the eccentricity of the ellipse that has the same second moments of the region considered at position one. The eccentricity is the proportion of the focal length (distance between the focal points) over the length of the main axis. The value must be in the range [0, 1]. When it is 0, the ellipse becomes a circle.

In addition to the features related to the geometry and intensities of the regions, for each one of these regions, were added attributes related to the patient's medical data informed by the physician at the time of its insertion. These features consist of a history of radiotherapy and chemotherapy treatment, history of cancerous diseases and bone lesions, this are represented with values of 1 and 0 symbolizing true or false.

At the end of the processing step, a set of feature vectors is obtained, each one referring to a region of high absorption detected in the image after the required processing and operations. For a better visualization of the obtained result, Fig. 6 outlines the enclosing rectangles and areas of each of the detected regions on the original image. Table 1 presents examples of some of the attributes that are present in the feature vectors built in this phase (Fig. 7).

Table 1. Examples of attributes collected for the detected regions.

Area	Perimeter	Ratio of areas	Mean intensity	Orientation	Eccentricity
138.00	61.70	0.77	0.49	0.16	0.81
108.00	53.32	0.75	0.32	0.93	0.81
150.00	72.28	0.79	0.33	−0.13	0.87

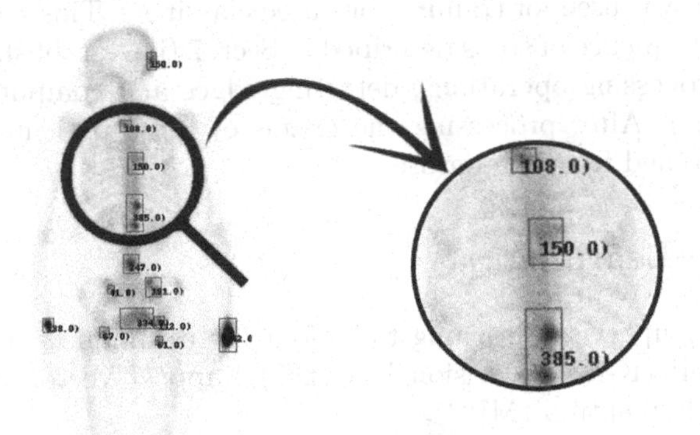

Fig. 6. Example highlighting detected regions and the areas obtained in mm^2 for each one.

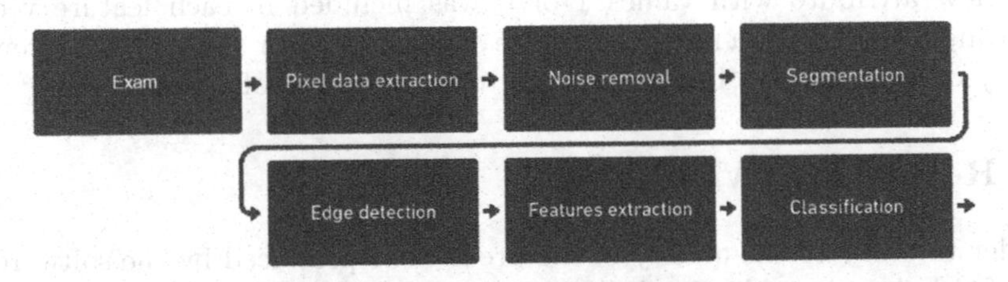

Fig. 7. Overview of processing steps.

8 Specification of Exams and Patients Database

As mentioned, the used database was provided by the nuclear physicians partners in this research. It is composed by 10 entire body scintigraphic exams, acquired according to the examination protocol and stored as DICOM standard. This research was approved by the Ethics Committee of the Brazilian Federal Government under the number CAAE: 45645121.7.0000.5243, title: "Optimization of a MACHINE LEARNING method for classification of scintigraphic images" in Portuguese: "Otimização usando método de MACHINE LEARNING para classificação de imagens cintilográficas" on 01.06.2021. Each exam is also accompanied by a detailed medical report, with the correct and accurate description (considered by medical consensus) of what was detected and the classification of each of the anomalies presented.

This dataset was used for training four used classifiers. This set was processed using the same sequence of steps described in Sect. 7 (i.e. establishing thresholds, doing image processing operations, detecting edges and computing features of detected regions). After processing the exams of these patients, a total of 77 entries was obtained from the dataset.

9 Machine Learning

To perform the supervised learning task, four (4) techniques were selected: K-Nearest neighbors (K-NN), Decision Tree (DT), Support Vector Machine (SVM) and Multilayer Perceptron (MLP).

Before applying these algorithms, the elements of the feature vectors were normalized to be in the [0, 1] range. For this, the entire 77 set of high absorption regions were used to find the maximum and minimum values of each of the 10 attributes that composes a vector position, for normalization, was applied the following equation:

$$x_{norm} = \frac{(x - x_{min})}{(x_{max} - x_{min})} \tag{1}$$

where (x_{max}) and (x_{min}) represent the maximum and minimum values of an specific x attribute, respectively.

A new attribute with values 1 or 0 was included in each feature vector, indicating respectively: a metastatic bone disease region or an area with intensive activity of radionuclides but not related to the presence of cancer cells.

10 Results and Validation

In order to evaluate the quality of the predictions produced by the software for each of the classifiers, the database was divided using the technique of stratified random division into training and test sets. For the training set, 70% of the samples contained in the database were used, while the remaining 30% were used to quantify the accuracy of the results.

For the KNN algorithm, was selected a number of neighbors equals to 1, since during the executions, this parameter presented the highest accuracy value. For the MLP, were taken as hyper-parameters 100 hidden layers, Adam optimizer and logistic activation function; SVM parameters were radial basis function kernel with a gamma of 0.08 and no hard limit on iterations.

Accuracy, recall, precision and the area under the receiver's operating characteristic curve (ROC) were computed with the obtained results. The results of the training and test sets are presented in Tables 2 and 3.

Table 2. Results for training set.

Classifier	Accuracy (%)	Precision (%)	Recall (%)	ROC
K-Nearest Neighbor	100	100	100	1.00
Multilayer Perceptron	81	84	57	0.76
Decision Tree	100	100	100	1.00
Support Vector Machine	94	100	84	0.92

Table 3. Results for test set.

Classifier	Accuracy (%)	Precision (%)	Recall (%)	ROC
K-Nearest Neighbor	82	70	87	0.83
Multilayer Perceptron	73	62	62	0.71
Decision Tree	86	77	87	0.87
Support Vector Machine	78	66	75	0.77

11 Conclusion and Future Works

Based on the results of the implemented learning techniques, it is possible to state that this is a very promising research, since despite having a low number of samples, the DT technique reached for all index values above 77%. The low results obtained by the SVM and MLP indicate the oblivious need of a larger set for training to improve these models.

Another aspect to be observed are the result of 100% accuracy obtained by the DT for the training set, suggesting a possible overfitting despite the relatively short tree (built with 11 nodes and depth 4). These values must be observed carefully in future executions and if necessary, implement pruning and growth limitation algorithms.

Aspects for improvement are: the segmentation and classification stages, the inclusion of a larger number of data to increase the training database, new strategies for filters and better adjustment of the hyper-parameters for learning.

Analyzing the results obtained by the developed research, although still initial, it shows itself to be considerably promising to its established goal, this being offer support to a qualified professional in the tasks of examining, detecting and classifying metastatic bone diseases through scintigraphic images.

References

1. Papandrianos, N., Papageorgiou, E., Anagnostis, A., Papageorgiou, K.: Bone metastasis classification using whole body images from prostate cancer patients based on convolutional neural networks application. PLoS ONE **15**(8), 0237213 (2020)
2. Hutton, B.F.: The origins of SPECT and SPECT/CT. Eur. J. Nucl. Med. Mol. Imaging **41**(1), 3–16 (2014)
3. Van den Wyngaert, T., et al.: The EANM practice guidelines for bone scintigraphy. Eur. J. Nucl. Med. Mol. Imaging **43**(9), 1723–1738 (2016)
4. Rieden, K.: Conventional imaging and computerized tomography in diagnosis of skeletal metastases. Radiologe **35**(1), 15–20 (1995)
5. Hamaoka, T., Madewell, J.E., Podoloff, D.A., Hortobagyi, G.N., Ueno, N.T.: Bone imaging in metastatic breast cancer. J. Clin. Oncol. **22**(14), 2942–2953 (2004)
6. Even-Sapir, E., Metser, U., Mishani, E., Lievshitz, G., Lerman, H., Leibovitch, I.: The detection of bone metastases in patients with high-risk prostate cancer: 99mTc-MDP Planar bone scintigraphy, single- and multi-field-of-view SPECT, 18F-fluoride PET, and 18F-fluoride PET/CT. J. Nucl. Med. **47**(2), 287–297 (2006)
7. Sadik, M., Jakobsson, D., Olofsson, F., Ohlsson, M., Suurkula, M., Edenbrandt, L.: A new computer-based decision-support system for the interpretation of bone scans. Nucl. Med. Commun. **27**(5), 417–423 (2006)
8. Nakajima, K., et al.: Enhanced diagnostic accuracy for quantitative bone scan using an artificial neural network system: a Japanese multi-center database project. Eur. J. Nucl. Med. Mol. Imag. Res. **3**(1), 1–9 (2013)
9. Belcher, L.: Convolutional neural networks for classification of prostate cancer metastases using bone scan images. Lund University (2017)
10. Suzuki, S.: Topological structural analysis of digitized binary images by border following. Comput. Vis. Graph. Image Process. **30**(1), 32–46 (1985)

Viewpoint Selection for Fibrous Structures in a Pre-operative Context: Application to Cranial Nerves Surrounding Skull Base Tumors

Méghane Decroocq[1], Morgane Des Ligneris[1], Timothée Jacquesson[2], and Carole Frindel[1](✉)

[1] Univ Lyon, INSA-Lyon, Université Claude Bernard Lyon 1, UJM-Saint Etienne, CNRS, Inserm, CREATIS UMR 5220, U1206, Lyon, France
carole.frindel@creatis.insa-lyon.fr

[2] Skull Base Multi-disciplinary Unit, Neurological Hospital Pierre Wertheimer, Hospices Civils de Lyon, 59 Bd Pinel, Lyon, France

Abstract. In this work, we present a viewpoint selection method specifically designed for fibrous structures in a pre-operative context. A view quality metric based on entropy was developed, which integrates the typical requirements of surgery planning. We applied our approach in the case of cranial nerves surrounding skull base tumors. The relevance of the viewpoints selected by our method was assessed qualitatively by a neurosurgeon and quantitatively based on statistical tests. These viewpoints were proven to have a high informative content, and therefore to enable a good understanding and mental representation the 3D anatomical scene in a pre-operative context.

Keywords: viewpoint selection · Entropy · Fiber tractography · Cranial nerves · Skull base tumor · Surgical planning

1 Introduction

Skull base tumor surgery remains a challenge since it requires complex surgical approaches reaching deep-seated tumors within a dense anatomical environment [11]. This environment includes cranial nerves, which are bundles of white matter fibers with sensorial or motor functions (e.g. the optic nerve). The preservation of the cranial nerves functions is one of the main stakes of tumor resection surgery. In this context, a thorough visualization of the nerves surrounding or displaced by the tumor could be of help for intervention planning, as attested by recent studies [2,5,13].

Advances in dMRI have used the unequal movement of water molecules along axons to reconstruct the 3 dimensional trajectory of the white matter fibers through tractography. However, tractography involves a complex multistep processing pipeline and is still difficult to apply to small-scale structures such as

G. Rozinaj and R. Vargic (Eds.): IWSSIP 2021, CCIS 1527, pp. 53–64, 2022.
https://doi.org/10.1007/978-3-030-96878-6_5

cranial nerves [6]. As a result, tractography datasets might be hard to visualize due to excessive amount of streamlines that are running in very different directions. Moreover, it moves away from the conventional radiological practice where datasets are visualized in 2 dimensions (2D). In this context, the selection of the viewpoints which best enhances the display of the important anatomical structures, here the tumor and nerves, as well as the whole scene, is valuable for surgery planning. It would reduce the complexity of the data, facilitate the understanding of the scene in 3 dimensions and guide the selection of the operative viewpoint.

To our knowledge, viewpoint selection for tractography fibers has never been investigated in the literature. This idea has been explored for other anatomical structures like organs, bones, vessels and tumors [3,9], but the proposed methods, based on geometrical criteria such as minimum distance or occlusion between objects, can hardly be applied to scattered and complex structures like tractography. Besides, one of the purposes of this work is to propose a local metric to prioritize the fibers according to their anatomical relevance. Prioritization of the structures of interest for medical applications was not introduced in any of the previous works.

Shannon's entropy quantifies the information content in a dataset, and is commonly used in the computer vision field to find informative viewpoints on meshes [1]. Moreover, it was shown to be an interesting measure to filter and visualize velocity streamlines [4], which have a nature close to tractography fibers. In this paper, we propose to use the local entropy of fibers as a selection metric to select the best viewpoints on a 3 dimensional scene including a tumor and surrounding nerves.

In Sect. 2, we give details on the acquisition of the medical images, the tractography pipeline employed and the anatomy of the cranial nerves of interest. In Sect. 3 we describe the application of Shannon entropy to tractography fibers and the calculation of the viewpoint quality score. Our validation strategy is explained. Finally, in the Sect. 4, the quality of the selected views is assessed both qualitatively and quantitatively, demonstrating the pertinence of the selection method.

2 Material

2.1 Cranial Nerves

The fibers of the white matter connects the different areas of the brain. Cranial nerves are organized bundles of white matter fibers with important sensorial or motor functions. Five cranial nerves or nerve groups located near the skull base were considered in this study: the optic nerve (Chiasma); the oculomotor nerve (III); the trigeminal nerve (V); the fascial and cochleo-vestibular nerves group (NF) and the mixed nerves groups (NM). Their mean diameter was estimated according to the known anatomy [7,10,16], and reported in Table 1.

Table 1. Estimated diameter of the studied nerves from the literature.

Nerve definition	Nerve abbreviation	Diameter (mm)
Optic	Chiasma	10
Oculomotor	III	5
Trigeminal	V	7
Facial and cochleo-vestibular	NF	3
Mixed	NM	2

2.2 Patients

Patient data (n = 31) used in this work is based on the study carried out between December 2015 and December 2017 in [5] (IRB Number 2015-A01113-46). Inclusion criteria were: skull base tumor; at least two cranial nerves in contact with the tumor; legal capacity; consent provided after fair information; 3T MRI data with dMRI acquisition. Exclusion criteria were: MR contraindications.

2.3 MRI Acquisition

A set of MR sequences were acquired in order to reconstruct the anatomical structures of interest. T1 post contrast weighted sequence and T2 steady state sequence are high resolution images($0.23 \times 0.23 \times 0.34$ mm), which were used as an anatomical reference. The segmentation of the tumors was made manually from the T1 sequence by a neurosurgeon. Diffusion images were acquired in order to compute the trajectories of fibers. It encodes the local diffusion of water molecules in 32 directions. This modality have a lower spatial resolution ($1.75 \times 1.75 \times 2$ mm). Distortions were corrected using the top-up and eddy tools of the FMRIB software library (FSL) software [12].

2.4 Tractography

Tractography is the method used to reconstruct the trajectory of the white matter fibers from the diffusion images. Figure 1 shows both the tractography reconstruction and a per-operative view for the oculomotor nerve. In this study, the tractography process was carried out from the acquired diffusion images using the Mrtrix3 software [14]. A brain mask was drawn to restrain the fiber reconstruction to the brain area. A spherical constrained deconvolution (6 spherical harmonic terms) was used to create a map of orientation distribution function (ODF) from the 32 directions of diffusion images. Cubic region-of-interest for the initialization of the tractography were designed by overlaying the ODF map on the T2-weighted MRI in order to identify the location of the cranial nerves with a great precision. A probabilistic tractography algorithm was used for the tracking of cranial nerves from the regions of interest [5]. The minimum fiber length required for the tracking was set to 10 mm and the number of fibers of each nerve to be reconstructed was set from 200 to 1000 according to the estimated nerve diameter. The output is a list of the spatial coordinates of the fibers.

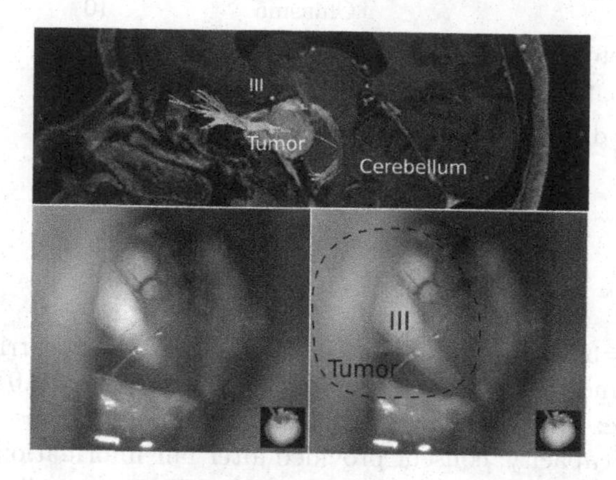

Fig. 1. Tractography fibers of the oculomotor nerve, superimposed on the T1 modality (top image). Post-operative view of the oculomotor nerve (bottom images). The position of the tumor before resection and the nerve trajectory are highlighted.

3 Methods

3.1 Entropy

Shannon's entropy is a measure commonly used in information theory, which quantifies the content of information in a dataset from its distribution. For a discrete random variable X with n classes, each class x_i having a probability $p(x_i)$ to appear, the entropy $e(X)$ is defined as:

$$e(X) = - \sum_{i=1...n} p(x_i) log_2(p(x_i)). \tag{1}$$

This measure can be easily applied to a vector field, by creating an orientation histogram of these vectors. With this orientation histogram, the probability of the vectors in the bin x_i, i.e. the vectors corresponding to a specific orientation, is calculated as:

$$p(x_i) = \frac{C(x_i)}{\sum_{i=1...n} C(x_i)}, \tag{2}$$

where $C(xi)$ is the number of vectors in bin x_i. Figure 2 illustrates this process in a two dimensional case for a case of orientation disorder (Fig. 2-a: low entropy) and of orientation coherence (Fig. 2-b: high entropy).

In a similar way, we can compute the entropy of the 3 dimensional vector field that encodes the local direction of the tractography fibers. In this way, entropy can be used to discriminate the fibers of homogeneous orientation, located in the core of the nerve, from the more chaotic fibers badly impacting the visual result.

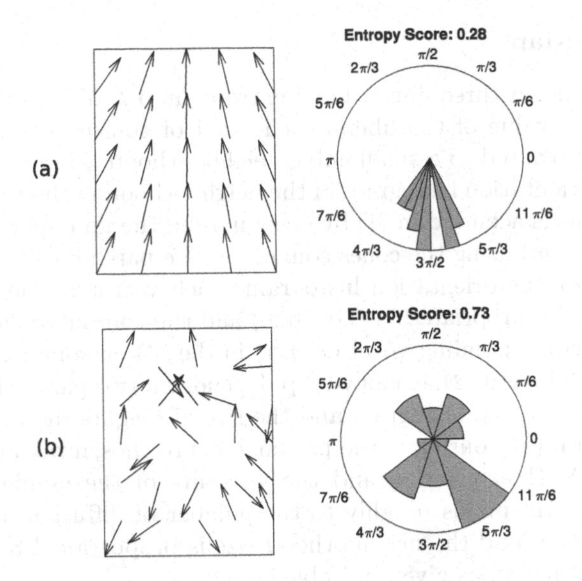

Fig. 2. Vector fields and their associated orientation histogram. The score given in the right column corresponds to the entropy value of the central vector calculated using a 3×3 neighborhood and 10 bins. The more the vector field is scattered the higher the entropy value.

A low entropy indicates that the location contains structures of medical interest and need to be preserved and enhanced in the visualization. The conversion of the fibers into a vector field is described in Sect. 3.2.

3.2 Vector Field Reconstruction

To apply entropy to tractography data, we first need to obtain a vector field of local fiber orientations. This information could be extracted from the main eigenvector of the diffusion tensor estimated from the raw dMRI data. However, this method is very sensitive to noise, particularly because of the low resolution of the dMRI data (2 mm) compared to the diameter of the nerves (2–10 mm: cf Table 1). We therefore propose to reconstruct a vector field from the fibers themselves. These are indeed less sensitive to noise because they have undergone numerous post-treatments during the of tractography process. Considering that fibers are sampled more than 10 times finer than the dMRI voxel, the resolution of the final vector field can be drastically improved.

In this sense, the fibers are first transformed into a 3D image encoding the local fiber density information. Then, a map of the maximum intensity gradient direction of this image is calculated using a $3 \times 3 \times 3$ neighborhood according to the method in [15]. Since the gradient orientation is normal to the actual fiber orientation, the vectors are reoriented according to the average of the vector products of the central voxel and its neighbors in the $3 \times 3 \times 3$ neighborhood.

3.3 Entropy Map

In order to produce a three dimensional entropy map $E(x, y, z)$ that represents the local entropy value of the fibers, each voxel of coordinates (x, y, z) in the vector field is associated to a small cubic neighbourhood $n \times n \times n$. Considering 3D vectors, an orientation histogram in the neighborhood of the considered voxel is computed. This is achieved in 3D by decomposing the unit sphere into patches of equal area [8], and using the cones connecting the patches to the center of the sphere as bins for the orientation histogram. Each vector in the neighborhood is assigned to the appropriate bin (i.e. cone) and the computed entropy value is assigned to the corresponding voxel (x, y, z) in the 3D entropy map $E(x, y, z)$.

Based on Eq. (1) and (2), entropy map depends on two parameters; the number of bins n used in the histogram and the size of neighborhood considered to build the histogram. In our case, the parameters are chosen taking into account priors on the dMRI acquisition and the anatomy of the cranial nerves. The number of bins corresponds roughly to the number of diffusion directions used in dMRI acquisition and the neighborhood size is proportional to the diameter of the considered nerve, as given in Table 1.

The resolution of the computed vector field (see Sect. 3.2) also impacts the computed 3D entropy map. As shown by Fig. 3, working at a better resolution makes it possible to use a neighborhood size for entropy computation smaller than the diameter of the nerve and hence to have a low entropy at the center of the nerve, as expected.

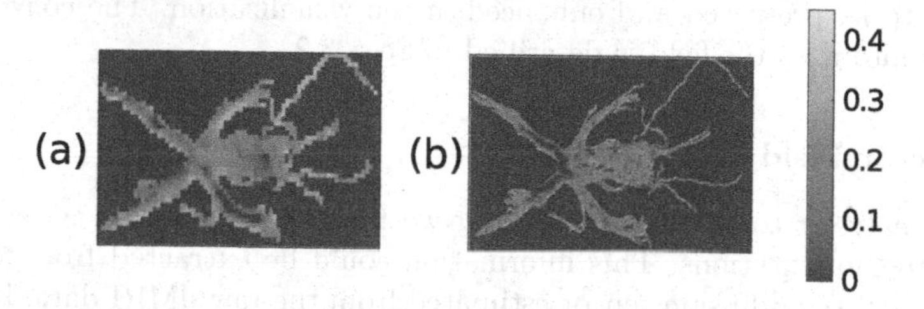

Fig. 3. Minimum entropy projections along the axial view for the optic nerve. Entropy maps were built from reconstructed vector fields of resolution 2 mm for (a) and 0.2 mm for (b).

3.4 Viewpoint Selection

Viewpoints are finally evaluated based on the information of the 3D entropy map. For a given viewpoint, we compute a 2D projection of the entropy map according to the specific view angle, as illustrated by Fig. 4. For each pixel of the 2D projection, we store the minimum entropy value found in the given direction. The projections are hereafter referred to as minimum entropy projection (MEP). The average of the entropy values in the MEP, which we call entropy score, is

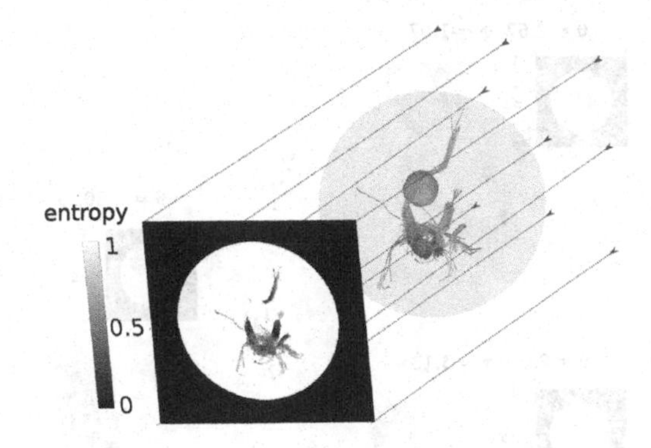

Fig. 4. Minimum entropy projection of a scene with a tumor and several nerves according to a given viewpoint. For clarity, a mesh of the tumor and fibers (colored according to their main direction) is represented instead of the 3D entropy map.

used as a quality metric for the viewpoint. A viewpoint with a low entropy score is considered to display relevant information about the 3D scene.

In order to combine the different anatomical structures which compose the scene, the raw entropy map has to undergo some pre-treatments before the projections. First, the entropy map of each cranial nerve is min-max normalized to give the same importance to each nerve in the scene to be visualized. Second, to prevent the occlusion of cranial nerves by themselves, the entropy is computed in a isotropic environment (in terms of number of pixels) to average the entropy on the same number of pixels regardless of the cranial nerve orientation. To do so, a bounding sphere of maximal entropy (entropy = 1) centered on the nerves is used.

Finally, the tumor-nerve occlusion is taken into account by including the tumor in the MEPs. A binary segmentation mask of the tumor is produced from the T2-weighted MRI data and registered to dMRI. The tumor voxels are then identified in the entropy map and set to the maximal entropy value (entropy = 1). As a result and illustrated in Fig. 4, the score of entropy of the MEP where the nerve hides the tumor and vice versa increases and the viewpoint associated is less likely to be selected.

For every scene, the MEPs associated 60 different angles in spherical coordinates (θ, ϕ) with $\theta \in [0, \pi]$ and $\phi \in [-\pi, \pi]$ are produced. A total of 60 view angles, equally distributed over the unit sphere according to [8], are evaluated. MEPs are ranked according to their entropy score. The viewpoints with the lowest and highest entropy scores are considered respectively best and worst viewpoints, as illustrated by Fig. 5 in the case of the optic nerve.

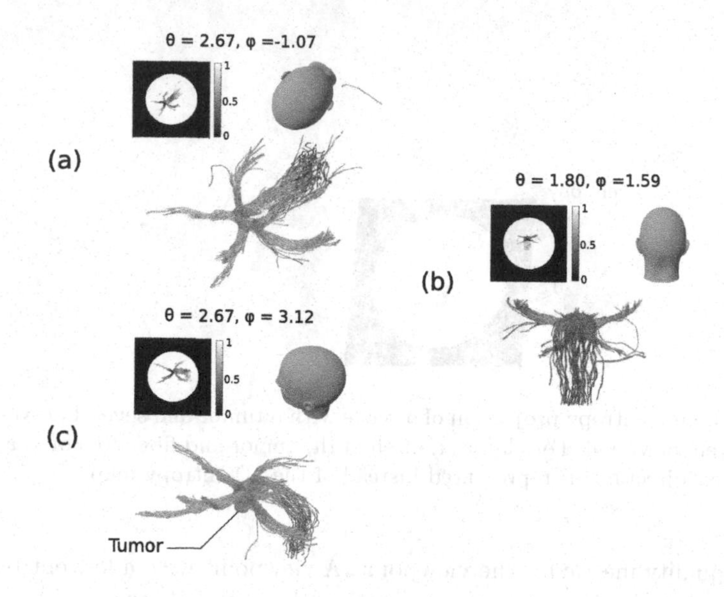

Fig. 5. Result of the best (a) and worst (b) viewpoints selection on the optic nerve. (c) shows the change in best viewpoint if we add a tumor (in purple). For each view, the corresponding MEP is given. (Color figure online)

3.5 Validation

A surgical intervention planning context is simulated in order to measure the usefulness of the viewpoint selection algorithm in clinical routine. For all the patients included in our study, the cranial nerves of surgical interest were identified by a neurosurgeon: they correspond to the nerves that are very close to the tumor and might be damaged during surgery. For each patient, the entropy score of 60 viewpoints on the tumor and nerves of surgical interest is computed, as described in Sect. 3.4. The viewpoints of minimal entropy score E_{min} and maximal entropy score E_{max}, referred hereafter respectively as best and worst viewpoint, were identified.

The performance of the viewpoint selection is first evaluated qualitatively by comparing the best viewpoint selected to the viewpoint chosen for surgery. The idea is to assess if the selected viewpoint can retrieve or surpass the surgical viewpoint. For this, we asked the neurosurgeon to systematically qualify it as better, equivalent or worse than the surgical viewpoint. In other words, the expert must assess whether the selected view better highlights the nervous structures in relation to the environment (other nerves and tumor) for the purpose of tumor resection.

The global performance of the proposed viewpoint selection method can be quantified from the appreciations given by the expert by computing the prevalence of better or equivalent views such as:

$$\text{prevalence} = \frac{|\text{Sup}| + |\text{Eq}|}{|\text{Sup}| + |\text{Eq}| + |\text{Inf}|}, \quad (3)$$

where Sup, Eq and Inf respectively correspond to the patient cases where the viewpoint associated with E_{min} is superior, equivalent or inferior to the surgical viewpoint according to the expert.

Finally, the entropy scores associated to respectively E_{min} and E_{max} and E_{min} and E_{surg} are compared on the basis of a paired sample t-test. The entropy E_{surg} associated with the surgical viewpoint was estimated from its orientation, which was repositioned in the framework of the 60 viewpoints tested in Sect. 3.4 and associated with the nearest viewpoint on the basis of a Euclidean distance on the angles.

4 Results

Counting the occurrences of binary score results at the scale of all patients, as explained in Sect. 3.5 enables us to assess that the viewpoint of best entropy, compared to the surgical viewpoint, provides additional information in 60% of the cases (17/28), a similar level of information in 28% of the cases (7/28) and bring a lower level of information in 14% of the cases (4/28). It was not possible for the neurosurgeon to assess three surgical cases, which is why the occurrence was performed on a total of 28 patients and not 31 as announced in Sect. 2.2. Overall, the view selection method provides additional or similar information compared to the surgical viewpoint up to 88% (see Eq. 3). Regarding the fact that the surgical viewpoint had been carefully selected by the medical expert, this results shows that the entropy score proposed in this work is relevant in a clinical context and meets the requirements of neurosurgeons in terms of visualization.

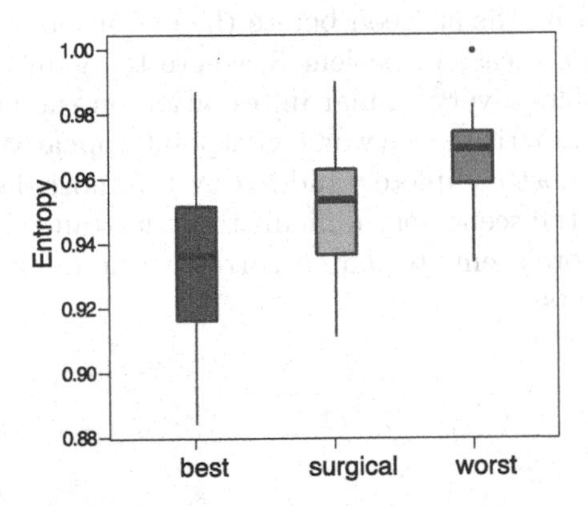

Fig. 6. Distribution of entropy associated with viewpoints in patients. The viewpoints associated to E_{min}, E_{max} and the surgery are respectively annotated "best", "worst" and "surgical".

Moreover, a statistical analysis of the entropy scores associated with the best, surgical and worst viewpoint was conducted as proposed in Sect. 3.5. The Fig. 6, illustrates the distribution of the entropy scores E_{min}, E_{surg} and E_{max} for all the patients. A paired sample t-tests shows a significant difference between the entropy scores of the best and the worst viewpoints (p-value = 3e−13) and between the entropy scores of the best and surgical viewpoints (p-value = 2e−06). The best viewpoints scores significantly better than the surgical viewpoint, which indicates that the surgical viewpoint selection can be improved by our algorithm.

Furthermore, as illustrated on Fig. 7, the best viewpoint selected with our algorithm clearly depicts the tumor and the displaced nerves. Although occlusion areas still exist, this viewpoint seems to provide the best trade-off between showing the nerves of interest in their entirety and minimizing occlusion with the tumor and the most disorganized fibers. Figure 7 shows the cases of three patients with respectively medium, high and low information gains from the surgical viewpoint as depicted by the curves on top of each row.

In the case of patient 1, the worst viewpoint is particularly unfavorable: high entropy fibers are present in the foreground, causing occlusion of both the tumor and the nerves. On the contrary, from the best viewpoint, the trajectory of the nerves V, III, and NF can be clearly observed, even if the nerve III and the tumor partially overlaps. The surgical viewpoint gives less information on the nerves and their context because the tumor masks nerves III and V and NF is hidden behind III, but still outperforms the worst view. The case of patient 2 is the most interesting as regards the contribution of our algorithm. Because of the important size of the tumor and the very close location of the nerves, very few viewpoint enables an optimal representation of the scene. Few choices were offered for the approach way of the tumor, resulting in a very poor visualization on the surgical viewpoint where the huge tumor masks almost completely the nerves NF and V. Even for this difficult case, our algorithm was able to find a viewpoint where all the nerve and their trajectory can be clearly identified. This representation can therefore be seen by thc surgeon before the operation and used for the pre-surgery planning. In the case of patient 3, where the gain is low, the best and surgical viewpoint offers a very similar information on the nerve trajectory and tumor position. In comparison, the worst viewpoint appear very poor: the tumor and nerve NF are almost completely hidden by the most disorganized fibers of the nerve V, making the scene very difficult to understand. In all the illustrated cases, the entropy score seems to match correctly the qualitative evaluation of the different viewpoints.

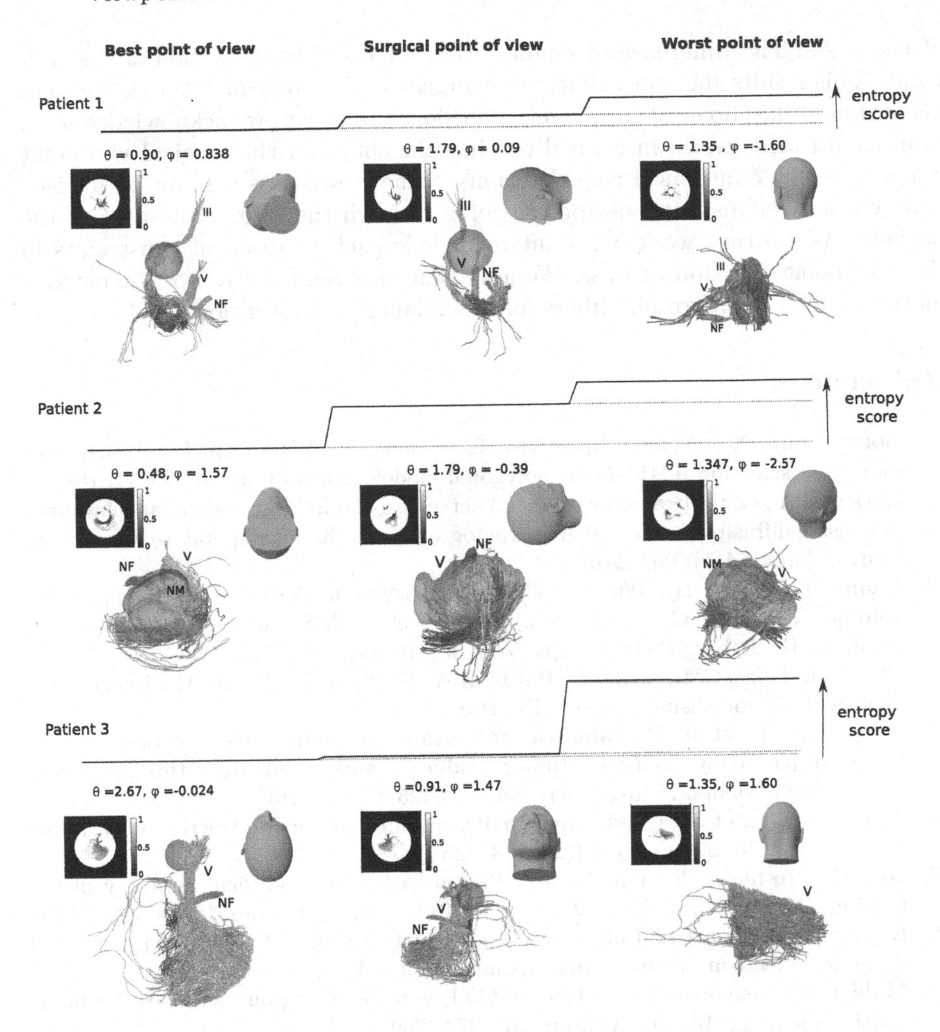

Fig. 7. MEPs, anatomical reference and scene visualization for the surgical viewpoint and the best and worst viewpoints returned by our algorithm. The difference of entropy score between the 3 viewpoints is given through a step function. A partial transparency of the tumor helps the visualization of the fibers inside and behind it.

5 Conclusion

In this paper, we presented a viewpoint selection framework for fibrous structures applied in the context of tumors surrounded by cranial nerves. The entropy of the direction of the white matter fibers was identified as an interesting metric to enhance the areas of the scene with high informative content (i.e. inside a nerve), in accordance with the concerns of the surgery. The occlusion caused by the tumor was taken into account. The best viewpoints selected by our algorithm were judged equivalent or superior than the viewpoint used for surgery in 88%

of the cases. The difference of quality score for those two viewpoints is significant. This results indicates that neurosurgeons could benefit from the present algorithm in the choice of the surgery viewpoint. However, we acknowledge some limitation to this work; in clinical routine, the choice of the surgical viewpoint is restricted by anatomical considerations. Some viewpoints can not be realistically chosen, for instance viewpoints going through the face or the neck of the patient. As a future work, we want to include such anatomical constraints in the viewpoint selection process. We also plan to extend the use of the entropy metric to filter tractography fibers for an enhanced visualization.

References

1. Bonaventura, X., Feixas, M., Sbert, M., Chuang, L., Wallraven, C.: A survey of viewpoint selection methods for polygonal models. Entropy **20**(5), 370 (2018)
2. Borkar, S.A., et al.: Prediction of facial nerve position in large vestibular schwannomas using diffusion tensor imaging tractography and its intraoperative correlation. Neurol. India **64**(5), 965 (2016)
3. Chan, M.-Y., Qu, H., Wu, Y., Zhou, H.: Viewpoint selection for angiographic volume. In: Bebis, G., et al. (eds.) ISVC 2006. LNCS, vol. 4291, pp. 528–537. Springer, Heidelberg (2006). https://doi.org/10.1007/11919476_53
4. Chen, M., Feixas, M., Viola, I., Bardera, A., Shen, H.W., Sbert, M.: Information Theory Tools for Visualization. CRC Press (2016)
5. Jacquesson, T., et al.: Probabilistic tractography to predict the position of cranial nerves displaced by skull base tumors: value for surgical strategy through a case series of 62 patients. Neurosurgery **85**(1), E125–E13 (2018)
6. Jacquesson, T., et al.: Overcoming challenges of cranial nerve tractography: a targeted review. Neurosurgery **84**(2), 313–325 (2018)
7. Joo, W., Yoshioka, F., Funaki, T., Rhoton, A.L.: Microsurgical anatomy of the trigeminal nerve. Clin. Anat. **27**, 61–88 (2014). https://doi.org/10.1002/ca.22330
8. Leopardi, P.: A partition of the unit sphere into regions of equal area and small diameter. Electron. Trans. Numer. Anal. **25**, 309–327 (2006)
9. Mühler, K., Neugebauer, M., Tietjen, C., Preim, B.: Viewpoint selection for intervention planning. In: EuroVis, pp. 267–274 (2007)
10. Rhoton, A.L.: Microsurgical anatomy of the posterior fossa cranial nerves. Clin. Neurosurg. **26**, 398–462 (1979)
11. Samii, M., Gerganov, V.M.: Petroclival meningiomas: quo vadis? World Neurosurg. **75**(3–4), 424 (2011)
12. Smith, S.M., et al.: Advances in functional and structural MR image analysis and implementation as FSL. Neuroimage **23**, S208–S219 (2004)
13. Song, F., et al.: In vivo visualization of the facial nerve in patients with acoustic neuroma using diffusion tensor imaging-based fiber tracking. J. Neurosurg. **125**(4), 787–794 (2016)
14. Tournier, J.D., et al.: *MRtrix*3: a fast, flexible and open software framework for medical image processing and visualisation. Neuroimage **202**, 116137 (2019)
15. Xu, C., Prince, J.L.: Snakes, shapes, and gradient vector flow. IEEE Trans. Image Process. **7**, 359–369 (1998)
16. Yoshino, M., et al.: Visualization of cranial nerves using high-definition fiber tractography. Neurosurgery **79**, 146–165 (2016). https://doi.org/10.1227/NEU.0000000000001241

Gait Recognition with DensePose Energy Images

Philipp Schwarz[1](✉), Josef Scharinger[2], and Philipp Hofer[3]

[1] LIT Secure and Correct Systems Lab, Johannes Kepler University, Linz, Austria
philipp.schwarz@jku.at

[2] Institute of Computational Perception, Johannes Kepler University, Linz, Austria
josef.scharinger@jku.at

[3] Institute of Networks and Security, Johannes Kepler University, Linz, Austria
philipp.hofer@ins.jku.at

Abstract. In the recent years, special emphasis has been placed on visual-based gait recognition due to its unique characteristics such as not requiring a special user action, or its long-distance recognizability. In general, there exist two methods - model-based and appearance-based methods - both of which come with their own advantages and disadvantages. In an effort to harness the best of both worlds we create a compact 3D human model-based gait representation out of 2D images with the help of the DensePose algorithm. We design a simple CNN and train several instances to show that the obtained gait representation can in fact be used to improve gait recognition accuracy. Experimental results are based on the publicly available CASIA-B dataset.

Keywords: Biometrics · Gait recognition · DensePose Energy Image

1 Introduction

Gait recognition describes the process of recognizing a person based on body shape or walking style. In contrast to other biometric modalities, gait possesses unique properties which make it an attractive alternative for recognition purposes. For instance, gait recognition does not require a certain user action such as entering a password, providing a fingerprint or looking into a camera. It does not require user cooperation or physical contact, which might be particularly of interest in turbulent times of COVID-19. Gait data is simple to collect and most notably gait is the only biometric modality which can be processed from very large distances. Furthermore, gait recognition can be performed with data from different kinds of sensors. A list of wearable sensors is provided in [7]. Apart

The research reported in this paper has been partly supported by the LIT Secure and Correct Systems Lab funded by the State of Upper Austria and by the Austrian Ministry for Transport, Innovation and Technology, the Federal Ministry of Science, Research and Economy, and the Province of Upper Austria in the frame of the COMET center SCCH.

G. Rozinaj and R. Vargic (Eds.): IWSSIP 2021, CCIS 1527, pp. 65–70, 2022.
https://doi.org/10.1007/978-3-030-96878-6_6

from those, gait data can also be acquired with audio sensors [8] or visual-based sensors such as cameras. This work focuses on the latter. Apart from the above mentioned advantages and the fact that gait data is very versatile regarding data acquisition, there are a few drawbacks. When it comes to visual-based gait data we can differentiate between internal and external factors that influence gait recognition. Internal factors include walking speed, weight change, age, mood, drugs, injuries, diseases, pregnancy or physical training. External factors include multi-gait conditions [1], clothing, shoes, walking surface, view-point, (self-)occlusion, shadows, reflections and accessories like bags. When it comes to visual-based gait recognition, two approaches can be found in literature namely model-based and appearance-based approaches. Model-based methods map the input to an underlying human-like skeleton model, while appearance-based methods compute feature vectors based on the raw input data. One advantage of model-based methods is that they can handle self-occlusion and differing viewing-angles better than appearance-based methods, especially 3D models [5]. This comes to the disadvantage of additional costs regarding data acquisition and computation. Contrary to this, appearance-based methods are cheap and simple to compute. Among the most common examples of appearance-based features is the Gait Energy Image (GEI).

In this work the main focus is to show that acquiring useful gait information from a 3D human model for identification purposes can effectively be done with machine learning approximations as provided by DensePose [2] and does not necessarily require an expensive camera setup. DensePose is a machine learning based system that is able to accurately map a 2D image of a person onto the surface map of a 3D human model. Visual-based gait recognition can be partitioned into the following sequence of processes:

a) **Object Recognition**: Check if a person is present in a frame.
b) **Object Detection**: Find all persons in a frame.
c) **Instance Segmentation**: Find all pixels that belong to one person.
d) **Object Tracking**: Track person over space and time.
e) **Feature Extraction**: Compute a feature vector for identification purposes.

Our work leverages the capabilities of DensePose which takes care of steps a) to c). Furthermore, object tracking is not relevant in this work, since experiments are conducted on data that only contains one person instance per frame. Since we want to show that the information obtained from DensePose is indeed useful for identification purposes, we choose a simple gait representation to conduct our experiments on, i.e. a variation of the GEI which we call DensePose Energy Image (DPEI).

The video data which is used in this work is taken from the publicly available dataset CASIA-B [9]. We apply the DensePose algorithm to all input frames to acquire 3D-like human silhouettes from the 2D images provided by CASIA-B. A simple convolutional neural network (CNN) is utilized to transform gait features like DPEIs and GEIs respectively into feature vectors which are used for classification. This approach potentially allows to enjoy the advantages of

3D model-based gait recognition as well as the rather low computational costs of 2D appearance-based methods.

2 Related Work

Since research on gait recognition has been done for decades, there already exists a plethora of gait representations. Appearance-based methods are based on a person's shape or silhouette. The best known example is the Gait Energy Image which was originally proposed in [3]. It is computed by averaging the centered silhouettes of a gait sequence, thereby representing a gait sequence compactly in a single image without the complete loss of temporal information.

While 3D model-based methods tend to be more accurate, they suffer from the problem that data acquisition is expensive. The approach in [5] for instance, requires multiple calibrated cameras to obtain gait information from their motion tracking algorithm.

The authors of [4] require a special camera that can record depth information. However, they also complement their data acquisition process by utilizing a human pose estimation algorithm.

The research in [6] goes one step further than DensePose by directly reconstructing the 3D shape of a person from a 2D image. Even though it only works well on high resolution images, this can potentially be used to overcome the viewing-angle challenge in gait recognition methods.

3 Proposed Method

3.1 CASIA-B Dataset

The CASIA-B dataset [9] contains walking sequences of 124 different subjects. Each walking sequence is captured by 11 cameras from 11 equidistant viewing angles. Each subject is recorded ten times. Six times under normal conditions, twice under carrying conditions and twice under changing clothing conditions. This dataset also contains binary person silhouettes for each video, but they are not needed in our work. From each of the 124 subjects four normal walks are used for training and the remaining two normal walks are used for testing.

3.2 Preprocessing

We preprocess each video by applying the DensePose algorithm to each video frame. What we get is another image where every pixel of the 2D input image that belongs to a person is mapped onto the texture map of a 3D human model. This means we get a silhouette-like representation where every pixel of the person is assigned a color value that corresponds to the UV-coordinates of the texture map space of the corresponding part of the 3D human model. Figure 1 shows an example of how this looks like. Unfortunately, in some cases the DensePose algorithm annotates parts of the background as a person too which is also illustrated in Fig. 1. To remove those artifacts we first extract the non-zero columns of an image. If the column indices do not form a consecutive chain, we can infer that there exist at least two person annotations. In this case, we delete all the columns that are not connected to the larger sequence of non-zero column indices. Overall, this seems to work pretty good as the artifacts are rather small compared to the real person silhouettes.

Fig. 1. Example input image from CASIA-B [9] and output of DensePose.

Additionally, the first and last ten frames of each video are discarded due to two different reasons. Either the person is too close to the camera which results in having the upper body cover most of the frame, or the person has not yet entered the frame completely. After excluding these 20 frames, sets of 25 consecutive frames are created with a step size of five. After manually inspecting the dataset we found that on average 25 frames make up a complete gait cycle. The step size of five is chosen as a trade-off between generating enough data for the training process and producing data with enough variation to counter overfitting.

In case there are still frames included that barely contain a person silhouette, i.e. frames with a silhouette pixel width below ten, a mechanism is in place to exclude them. As mentioned above, this can occur when a person enters/leaves the frame and potentially reduces the number of frames per Energy Image below 25, however this should happen only rarely. Only in four cases (subjects 037, 051, 109, and 120) we ended up with not enough frames after preprocessing. As this is the first study in which we want to evaluate the potential of our approach, we simply excluded the whole subject from the experiment. This does not influence the evaluation as we exclude the subjects for all experiments equally.

3.3 DensePose Energy Image (DPEI)

We compute the DPEI analogously to the GEI. For each set of frames a DPEI is generated as follows. First, the average DensePose silhouette height is computed. Each DensePose silhouette is then scaled according to the average height and positioned in the center of the image. Summing up the pixel values and dividing them by the number of frames yields the final DensePose Energy Image. GEIs do not need to be computed separately, we just convert the DPEIs to gray-scale.

3.4 Network Architecture

The GEIs as well as the DPEIs introduced in this work represent the input of a simple CNN. The architecture of the CNN as well as number of filters and filter size is visualized in Fig. 2. We choose Stochastic Gradient Descent with a learning rate of 0.001 to optimize the categorical crossentropy loss between labels (subject IDs) and predictions. Design decisions are based on empirical tests, without spending much time on fine tuning.

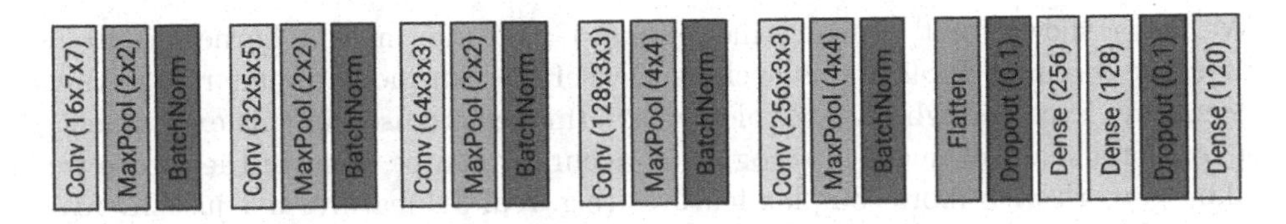

Fig. 2. Neural network architecture

4 Experimental Results

We conducted three experiments. In experiment 1 the CNN described in Sect. 3.4 is trained on the GEIs. In experiment 2 the same CNN architecture is used, with the exception of the color channels. Since DPEIs are RGB-images the CNN requires three color channels for training, whereas training on GEIs only requires one channel. Experiment 3 uses the exact same model as experiment 1, this time however we train not one but three separate models, one for each color channel of the DPEIs. The final accuracy score is computed as a weighted average of each of the three models.

The evaluation procedure for all experiments is the same. From the CASIA-B dataset only the videos recorded under normal walking conditions are used. After preprocessing we are left with 120 different subjects. The CNN performs a $1:N$ comparison where $N = 120$. We use 6-fold cross-validation over all six normal walks of CASIA-B, with each fold containing four walks for training and two walks for testing. This can be seen in Table 1 which also shows the Top-1 accuracy for each experiment as well as the average accuracy over all folds. The results of these initial experiments underline the potential of our approach. Compared to the classical GEIs, our DPEIs achieve slightly higher classification accuracies in experiment 2 and significantly higher accuracies in experiment 3.

Table 1. Classification accuracy per fold - highest accuracy is marked in bold.

Cross-validation split		Top-1 Accuracy			Top-5 Accuracy		
Train walks	Test walks	Exp. 1	Exp. 2	Exp. 3	Exp. 1	Exp. 2	Exp. 3
['01', '02', '03', '04']	['05', '06']	0.887	0.898	**0.933**	0.967	0.958	**0.969**
['02', '03', '04', '05']	['06', '01']	0.912	0.923	**0.952**	0.960	0.947	**0.969**
['03', '04', '05', '06']	['01', '02']	0.921	0.906	**0.955**	0.955	0.981	**0.982**
['04', '05', '06', '01']	['02', '03']	0.930	0.941	**0.959**	0.925	0.970	**0.986**
['05', '06', '01', '02']	['03', '04']	0.912	0.946	**0.953**	0.978	0.988	**0.994**
['06', '01', '02', '03']	['04', '05']	0.917	0.933	**0.957**	0.976	0.966	**0.990**
Average		0.913	0.924	**0.951**	0.960	0.968	**0.982**

5 Conclusion and Future Work

We have shown that utilizing model-based gait information harnessed from machine learning applications such as DensePose can increase gait recognition accuracy. Even though only simple Energy Images are used in this experiment, a consistent improvement in recognition accuracy can be seen. In the future we plan to work with more complex features (e.g. real 3D features not just 2D features that are derived from a 3D model) in an effort to overcome some of the challenges concerning gait recognition. For instance the viewing-angle problem could be interesting in this regard or gait recognition under clothing variations.

References

1. Chen, X., Weng, J., Lu, W., Xu, J.: Multi-gait recognition based on attribute discovery. IEEE Trans. Pattern Anal. Mach. Intell. **40**(7), 1697–1710 (2018)
2. Güler, R.A., Neverova, N., Kokkinos, I.: DensePose: dense human pose estimation in the wild. In: Proceedings of the IEEE Computer Society Conference on Computer Vision and Pattern Recognition, pp. 7297–7306 (2018)
3. Han, J., Bhanu, B.: Individual recognition using gait energy image. IEEE Trans. Pattern Anal. Mach. Intell. **28**(2), 316–322 (2006)
4. Huynh-The, T., Hua, C.H., Tu, N.A., Kim, D.S.: Learning 3D spatiotemporal gait feature by convolutional network for person identification. Neurocomputing **397**, 192–202 (2020)
5. Krzeszowski, T., Michalczuk, A., Kwolek, B., Switonski, A., Josinski, H.: Gait recognition based on marker-less 3D motion capture. In: 2013 10th IEEE International Conference on Advanced Video and Signal Based Surveillance, AVSS 2013, pp. 232–237 (2013)
6. Saito, S., Huang, Z., Natsume, R., Morishima, S., Li, H., Kanazawa, A.: PIFu: pixel-aligned implicit function for high-resolution clothed human digitization. In: Proceedings of the IEEE International Conference on Computer Vision, October 2019, pp. 2304–2314 (2019)
7. Wan, C., Wang, L., Phoha, V.V.: A survey on gait recognition. ACM Comput. Surv. **51**(5), 1–35 (2018)
8. Xu, W., Yu, Z., Wang, Z., Guo, B., Han, Q.: AcousticID. Proc. ACM Interact. Mobile Wearable Ubiquit. Technol. **3**(3), 1–25 (2019)
9. Yu, S., Tan, D., Tan, T.: A framework for evaluating the effect of view angle, clothing and carrying condition on gait recognition. In: Proceedings - International Conference on Pattern Recognition, vol. 4, pp. 441–444 (2006)

Adaptive IIR Filtering for System Identification Applying the Method by Nelder and Mead

Vassil Guliashki[1(✉)] and Galia Marinova[2]

[1] Institute of Information and Communication Technologies, Bulgarian Academy of Sciences, Acad. G. Bonchev Street 2, Sofia 1113, Bulgaria

[2] Technical University of Sofia, Kl. Ohridski Blvd. 8, Sofia 1000, Bulgaria
gim@tu-sofia.bg

Abstract. In this paper, a simple IIR filter is used in system identification. Uniform white sequence is used as an input signal for the unknown system. A noise white sequence signal, which is not correlated with the input signal, is added to the system output. An illustrative example is solved, and the optimization is performed using the Simplex method by Nelder and Mead. A comparison is done to results by a genetic algorithm and a simulated annealing algorithm. It is demonstrated, that a gradient based algorithm gets stuck in a local minimum. The obtained result confirms the efficiency and efficacy of this approach.

Keywords: System identification · Adaptive filtering · Convex optimization

1 Introduction

System identification is a broad field of research related to signal processing (see for example [1–3]), and connected with the use of various optimization techniques, such as Genetic algorithms (GA) [4, 5], Particle swarm optimization (PSO) algorithms [6, 7], Artificial been colony algorithms (ABC) [8], Neural networks (NN) [9, 10], and adaptive filtering using a least squares algorithm (LMS) on FIR and IIR adaptive filters [11–14]. Applications with adaptive IIR filters are discussed in [15].

For the purpose of finding the optimal parameters of the system, the adaptive filtering technique is used, in which the parameters of the filter change at each iteration. Adaptive filters are divided into filters with finite impulse response (FIR) and filters with infinite impulse response (IIR). It has been found that with the same number of coefficients IIR filters have much better performance than FIR filters [16, 17]. In addition to this advantage, IIR filters have two important shortcomings: 1) When optimizing the coefficients of the IIR filter, the objective function (error function surface) can be multimodal, i.e. the search for a global minimum is necessary. 2) During the adaptation process, the IIR filter may become unstable [18, 19]. The second shortcoming can be overcome by limiting the parameter space to an appropriate range of values. The first shortcoming is related to the fact that with a multimodal objective function, gradient-based methods usually finish the search process into local minima and cannot reach the global minimum. For this reason, metaheuristics and global optimization algorithms such as Tabu

G. Rozinaj and R. Vargic (Eds.): IWSSIP 2021, CCIS 1527, pp. 71–81, 2022.
https://doi.org/10.1007/978-3-030-96878-6_7

search (TS), Simulated annealing (SA), Differential evolution (DE), Genetic algorithms (GA), Particle swarm optimization (PSO), artificial bee colony (ABC) algorithms, and other evolutionary algorithms are used [17, 20–24]. In addition to these algorithms, the derivative-free method by Nelder and Mead [25] can also be used successfully.

A Genetic algorithm is used in [16] for multi-modal error optimization. Its negative features are the high computational complexity and the slow convergence. A hybrid LMS-GA having better performance is proposed in [2]. It is characterized by a simple implementation, less sensitivity to the parameters selection, ability to find out the global optimal solution, and faster convergence. The slow convergence and the poor efficacy are the main drawbacks of many evolutionary algorithms, as well as of SA algorithm. TS metaheuristic can have higher accuracy but its implementation is more complex.

In this paper, the system identification is realized through the technique of adaptive filtering, and the optimization is performed by means of a solver implementing the Nelder and Mead's simplex method. A solver based on a genetic algorithm and a solver based on the simulated annealing algorithm were used for comparison. Finding a local optimum using a gradient-based method is also shown. The obtained result confirms the advantages of the used approach.

The paper is organized as follows: Sect. 2 presents the problem formulation and the algorithm for its solution. An illustrative example is described in Sect. 3. The experimental results are presented in Sect. 4. Finally, the results obtained are discussed in the Conclusion.

2 Problem Formulation and Algorithm for Solving

2.1 Problem Formulation

The IIR filter equation has the form:

$$y(k) = \sum_{i=0}^{n} a_i x(k-i) - \sum_{j=1}^{m} b_j y(k-j) \tag{1}$$

where $y(k)$ is the output signal at time step k, and $x(k)$ is the input signal at time step k. Respectively $x(k-i)$ and $y(k-j)$ denote the input and the output signals i and j steps before the time step k. The coefficients $a_0, a_1, \ldots, a_n$; and $b_1, \ldots, b_m$; are parameters that have to be calculated. By m ($\geq n$) is denoted the filter order.

The transfer function is the relationship between the input and output signals. For discrete signals Z-transform is used and the IIR filter transfer function is expressed by:

$$\boldsymbol{H}_f(z) = \frac{A(z)}{B(z)} = \frac{a_0 + a_1 z^{-1} + \ldots + a_n z^{-n}}{1 + b_1 z^{-1} + \ldots + b_m z^{-m}}, \tag{2}$$

The System identification setup is shown in Fig. 1. The unknown system (unknown plant) has a transfer function $\boldsymbol{H_s(z)}$ and an adaptive filtering algorithm is applied to calculate the coefficients of IIR filter used to model the system.

The transfer function of the unknown system is expressed in a similar way:

$$\boldsymbol{H}_s(z) = \frac{S(z)}{V(z)} = \frac{s_0 + s_1 z^{-1} + \ldots s_{ns} z^{-ns}}{1 + v_1 z^{-1} + \ldots v_{ms} z^{-ms}}, \tag{3}$$

The coefficients $s_0, s_1, \ldots, s_{ns}$; and $v_1, \ldots, v_{ms}$; are system parameters that have to be calculated during the optimization process. By ms ($\geq ns$) is denoted the system order.

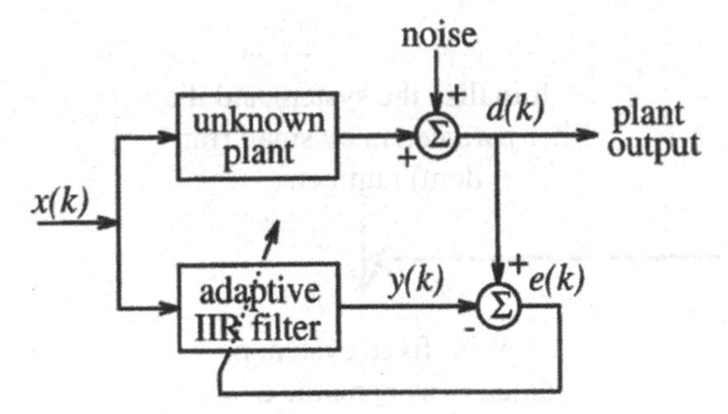

Fig. 1. Setup of adaptive IIR filtering for system identification

Let the filter coefficient vector is denoted by

$$w_f = [\mathbf{a}^\mathbf{T}\mathbf{b}^\mathbf{T}]^\mathrm{T} = [a_0, a_1, \ldots, a_n, -b_1, \ldots - b_m]^\mathrm{T} \tag{4}$$

and the system coefficient vector is denoted by

$$w_s = [\mathbf{s}^\mathbf{T}\mathbf{v}^\mathbf{T}]^\mathrm{T} = [s_0, s_1, \ldots, s_{ns}, -v_1, \ldots - v_{ms}]^\mathrm{T} \tag{5}$$

If the signal vector is denoted by $\boldsymbol{\varphi}^\mathbf{T}(k) = [x(k),\ldots, x(k-n), y(k-1),\ldots, y(k-m)]$, Eq. (1) can be written in the form of a linear regression:

$$y(k) = \mathbf{w}_f^\mathbf{T}(k) \cdot \varphi(k) \tag{6}$$

Let the noise signal be denoted by $p(k)$ and the system output - by $y_s(k)$. In this paper an output error formulation is considered. The error signal $e(k)$ has the form:

$$e(k) = d(k) - y(k), \tag{7}$$

where

$$d(k) = y_s(k) + p(k). \tag{8}$$

Alternative error formulations are discussed in [2, 26].

Here is formulated a time-averaged objective function

$$f(w_s) = \frac{1}{N}\sum_{k=1}^{N}(d(k) - y(k))^2 \tag{9}$$

for the system identification with adaptive filtering using the above setup. By N is denoted the data length (the number of time steps). Here $f(w_s)$ has to be minimized to obtain the optimal system parameters w_s. When the filter order m is smaller than the system order ms, the optimization problem can become multimodal [26].

Based on this formulation an adaptive filtering algorithm, similar to that one in [2] is used. The filter coefficients are adapted (updated consecutively) and for each fixed w_f vector the system parameter vector w_s is optimized minimizing the objective function (9).

2.2 Flowchart of the System Parameters Optimization Algorithm

The flowchart of adaptive filtering algorithm is presented in Fig. 2:

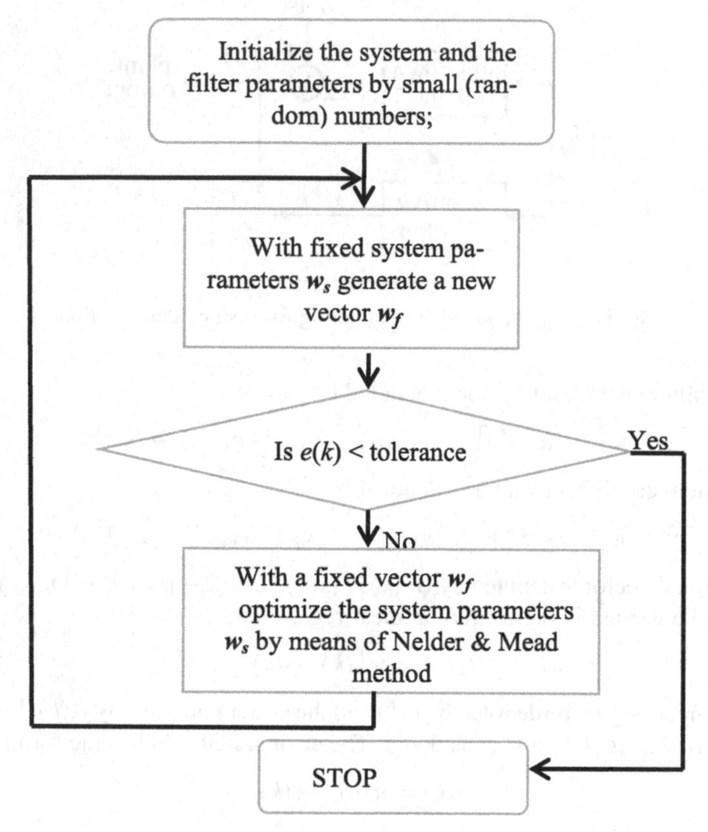

Fig. 2. Flowchart of adaptive filtering algorithm

2.3 Filter Coefficients Adaptation Algorithm

To update the filter coefficients a procedure similar to that one, described in [26] is used. The formula for filter coefficients updating is:

$$w_f(k+1) = w_f(k) + \alpha F^{-1}(k+1) \cdot \varphi^{\mathbf{T}}(k) \cdot e(k), \tag{10}$$

which is known as "recursive Gauss-Newton algorithm" [26, 27]. Here F is the estimate of the Hessian matrix of $f(w_f)$. The positive scalar α is the step size which controls the algorithm convergence rate. Also the so called "forgetting factor" $\lambda = 1 - \alpha$ is used in the formula for updating the estimate of Hessian matrix:

$$F(k+1) = \lambda F(k)\alpha \cdot \varphi(k) \cdot \varphi^{\mathbf{T}}(k). \tag{11}$$

It is assumed that $\lambda \in [0.9\ 1)$. The algorithm is initialized setting $\boldsymbol{F}(0)$ to be identity matrix: $\boldsymbol{F}(0) = \boldsymbol{I}$. In this study is assumed that $\lambda = 0.9$; $\alpha = 0.1$. The step size should be chosen sufficiently small, so that the filter coefficients adapt slowly. The algorithm for filter coefficients adaptation will converge if $\boldsymbol{F}$ is always positive definite and if the filter is stable, i.e. the poles of $\boldsymbol{B}(z)$ always should lie inside the unit circle [26].

3 Illustrative Example

To illustrate the system identification by the adaptive filtering the following example corresponding to the setup in Fig. 1 is used:

$$\boldsymbol{H}_s(\mathbf{z}) = \frac{s_0 + s_1 z^{-1}}{1 + v_1 z^{-1} + v_2 z^{-2}}, \tag{12}$$

and

$$\boldsymbol{H}_f(\mathbf{z}) = \frac{a_0}{1 + b_1 z^{-1}} \tag{13}$$

The system input signal is a uniform white sequence having values in the interval [0, 1]. The noise is also a white sequence, not correlated with the input signal and the SNR is 40 dB.

The data length used to calculate (9) is $N = 1000$. The filter order is smaller than the system order and the objective function is multimodal. Applying the above adaptive filtering algorithm the global optimal solution was found.

4 Test Results

The formulated optimization problem was solved initially by means of a Genetic algorithm with standard default options in MATLAB "Optimization toolbox", and with 2000 generations.

The initial filter parameters were:

$$a_0 = -0.4239; \quad b_1 = -0.8506;$$

The best obtained solution after 10 runs of "**ga**" solver is:

$$f(\boldsymbol{w}_s) = 0.22786590139436105\ \mathrm{E} - 3.$$

$$s_0 = -0.1291764026606757, \quad s_1 = -0.1857579753212366,$$

$$v_1 = -1.1632084380302095, \quad v_2 = 0.2699488019212094;$$

Then the problem was solved by the Simulated annealing algorithm in MATLAB "Optimization toolbox", starting with the same filter values.

The "**simulannealbnd**" solver obtained the following solution after 2000 iterations:

$$f(w_s) = 0.3859736161248466 \text{ E} - 3$$

$$s_0 = 0.0500009309144380, \; s_1 = -0.4761473455691334,$$

$$v_1 = 1.1378140941588529, \; v_2 = 0.2574377307374864;$$

Then lower and upper bounds on the system parameters were imposed, and the problem was solved by a gradient-based method (Interior point method) by "**fmincon**" solver.

The lower and upper bounds were:

$$\text{LB} = [-0.5, \; -1., \; -1.5, \; 0.], \quad \text{UB} = [1. \; 0. \; 0. \; 1.];$$

In 38 iterations the following solution was obtained:

$$f(w_s) = 0.21969251311102905 \text{ E} - 3$$

$$s_0 = 9.305345179303844E - 9, \quad s_1 = -0.1899447651947604,$$

$$v_1 = -0.2288134708171158, \quad v_2 = 0.07538157888894959;$$

This solution was locally optimal.

Then the problem was initially solved by the Nelder & Mead method [25] ("**fminsearch**" solver) without any constraints. After 384 iterations the following solution was obtained:

$$f(w_s) = 0.2054162721878904 \text{ E} - 3$$

$$s_0 = -0.10766247565753506, \quad s_1 = -0.3373361559811813,$$

$$v_1 = -0.8373482273289539, \quad v_2 = -0.0118407567339429;$$

The search process is shown in Fig. 3:

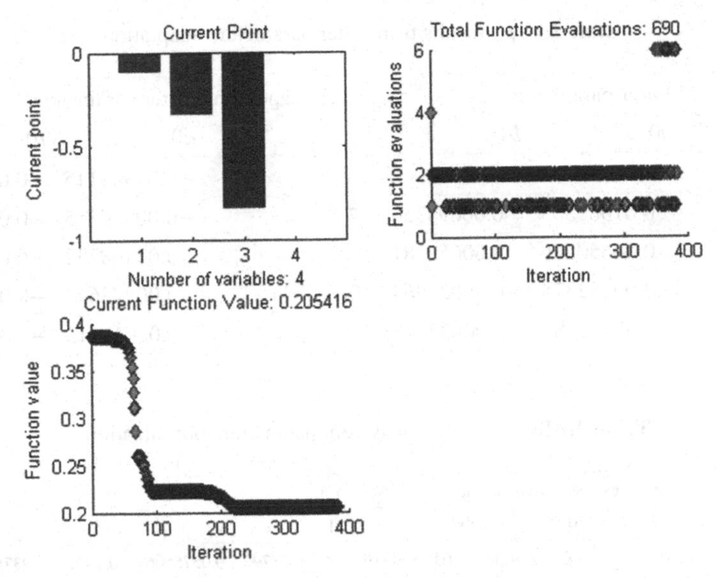

Fig. 3. Initial optimization of system parameters by means of Nelder & Mead method

The results from initial optimization by different solvers are presented in Table 1.

Table 1. Test results from initial optimization by 4 different solvers

Solver	$f(w_s)$	Iterations	System parameters	
			s_0, s_1	v_1, v_2
ga	0.22786590 E−3	2000	$s_0 = -0.129176$ $s_1 = -0.185758$	$v_1 = -1.163208$ $v_2 = 0.269949$
simulannealbnd	0.38597362 E−3	2000	$s_0 = 0.050001$ $s_1 = -0.476147$	$v_1 = 1.137814$ $v_2 = 0.257438$
fmincon	0.21969251 E−3	38	$s_0 = 9.3053$ E−9 $s_1 = -0.189945$	$v_1 = -0.228813$ $v_2 = 0.075382$
fminsearch	0.20541627 E−3	384	$s_0 = -0.107662$ $s_1 = -0.337336$	$v_1 = -0.837348$ $v_2 = -0.011841$

After that the 10 **consecutive** times of filter parameter adaptation and optimization of the system parameters, the system parameters were optimized. The obtained results are summarized in Table 2 and Table 3.

Table 2. Test results from filter parameters adaptation

Filter adaptation №	Filter parameters		Filter adaptation №	Filter parameters	
	$a0$	$b1$		$a0$	$b1$
1	0.07450850	0.00000000	6	−0.00903115	−0.00065132
2	0.01081365	0.00093356	7	−0.00930555	−0.00065463
3	0.00650676	0.00052281	8	−0.00968732	−0.00066194
4	0.00252375	0.00021969	9	−0.00998623	−0.00067347
5	−0.0035148	−0.00022428	10	−0.01012415	−0.00068159

Table 3. Test results from system parameters optimization

System optim. №	Iterations	System parameters				$f(w_s)$
		$s0$	$s1$	$v1$	$v2$	
1	467	0.01159215	−0.03564219	−0.01078790	0.05755069	0.2051056093787718 E−3
2	392	−0.06404563	−0.03707844	0.06288819	0.06160814	0.2049287784228326 E−3
3	294	−0.06841385	−0.03751504	0.06716681	0.06200449	0.2049205843211464 E−3
4	223	−0.07245369	−0.03795820	0.07112433	0.06238915	0.2049131296288906 E−3
5	234	−0.07857874	−0.03870096	0.07712497	0.06300324	0.2049020432504497 E−3
6	259	−0.08417464	−0.03945466	0.08260678	0.06359603	0.2048921312576315 E−3
7	223	−0.08445302	−0.03949383	0.08287945	0.06362616	0.2048916438430020 E−3
8	202	−0.08484032	−0.03954866	0.08325881	0.06366820	0.2048909664998534 E−3
9	201	−0.08514356	−0.03959188	0.08355583	0.06370122	0.2048904367127861 E−3
10	177	−0.08528349	−0.03961191	0.08369286	0.06371654	0.2048901923928147 E−3

The final error was smaller than the given tolerance.

The last two optimizations of the system parameters are shown in Fig. 4 and Fig. 5.

The last obtained output error values are very close and the correspondent error gradient is very small.

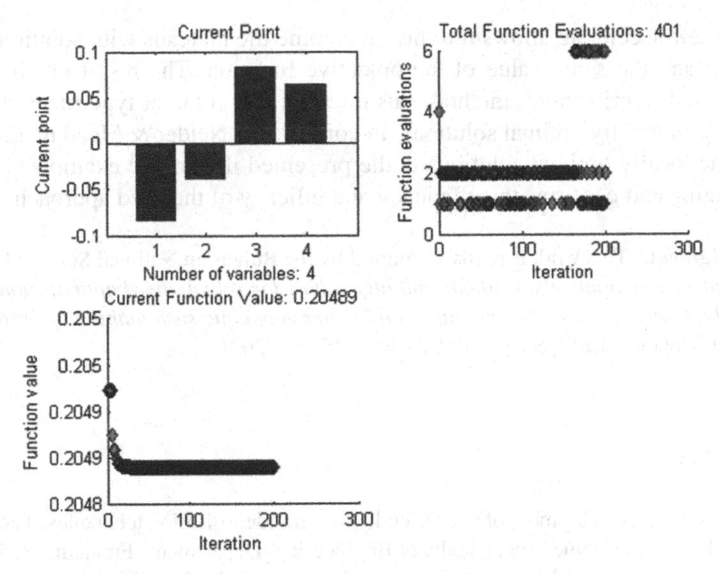

Fig. 4. Optimization of system parameters № 9

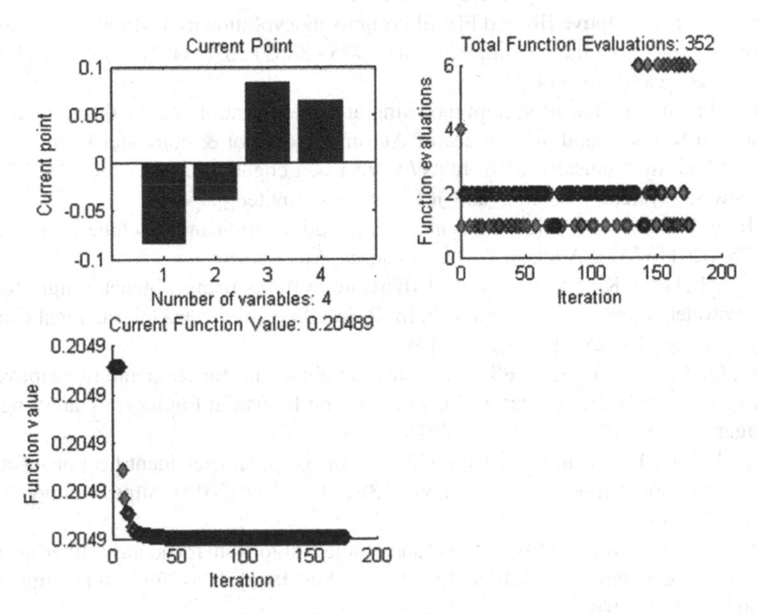

Fig. 5. Final optimization of system parameters

5 Conclusions

In this paper the system identification is realized through adaptive filtering, and the system parameters optimization is performed by means of the Simplex method by Nelder & Mead. The initial results, obtained by a Genetic algorithm and by a Simulated annealing show that the performance of the Nelder & Mead method is better. The property of this method to self-accelerate allows it to fast overcome the plateaus with solutions, having almost one and the same value of the objective function. The result obtained by the gradient based Interior point method illustrate that the gradient type methods can fall into the trap of locally optimal solutions. In contrast the Nelder & Mead method is able to avoid the locally optimal solutions in the presented illustrative example. This result is encouraging and confirms the efficiency and efficacy of the used approach.

Acknowledgment. This work is partly supported by the Bulgarian National Science Fund – the project "*Mathematical models, methods and algorithms for solving hard optimization problems to achieve high security in communications and better economic sustainability*", Grant No: KP-06-N52/7 and by the CEEPUS network CIII-BG-1103-05-2021.

References

1. Geerardyn, E.: Development of user-friendly aystem identification technques. Thesis for the degree Doctor in Engineering, Faculty of Engineering, Department "Fundamental Electricity and Instrumentation (ELEC)", Vrije University Brussels/VUB, ISBN: 978-9-4619743-2-7 (2016). https://github.com/egeerardyn/phdthesis
2. Ibraheem, I.K.: Adaptive IIR and FIR filtering using evolutionary LMS algorithm in view of system identification. Int. J. Comput. Appl. (0975 - 8887) **182**(11), 31–39 (2018). https://doi.org/10.5120/ijca2018917740
3. Favier, G.: An overview of system modeling and identification. In: 11-th International conference on Sciences and Techniques of Automatic control & computer engineering (STA 2010), Monastir, Tunisia (2010). https://www.researchgate.net/publication/266170681An_overview_of_system_modeling_and_identification_Invited_paper
4. Goldberg, D.E.: Genetic Algorithms in Search, Optimization and Machine Learning, ISBN-13: 978–0201157673. Addison Wesley, Reading, Mass (1989)
5. Titel, F., Belarbi, K.: Identification of dynamic systems using a genetic algorithm-based fuzzy wavelet neural network approach. In: Proceedings of the 3rd International Conference on Systems and Control, pp. 6–11 (2013)
6. Zhang, Q., Li, Q., Wu, S.: A PSO identification algorithm for temperature adaptive adjustment system. In: IEEE International Conference on Industrial Engineering and Engineering Management (IEEM), pp. 752–755 (2016)
7. Zhang, J., Xia, P.: An improved PSO algorithm for parameter identification of nonlinear dynamic hysteretic models. J. Sound Vibr. **389**, 153–167 (2017). https://doi.org/10.1016/j.jsv.2016.11.006
8. Karaboğa, N., Çetinkaya, M.B.: A novel and efficient algorithm for adaptive filtering: artificial bee colony algorithm. Turk. J. Elec. Eng. Comp. Sci. **19**(1), 175–190 (2011). https://doi.org/10.3906/elk-0912-344
9. Santosh Kumar Behera, D.R.: System Identification Using Recurrent Neural Network. Int. J. Adv. Res. Electr. Electron. Instrum. Eng. ISSN (Print): 2320–3765 **3**(3), 8111–8117 (2014)
10. Ibraheem, I.K.: System identification of thermal process using elman neural networks with no prior knowledge of system dynamics. Int. J. Comput. Appl. (0975-8887) **161**(11), 38–46 (2017)
11. Abadi, M.S.E., Mesgarani, H., Khademiyan, S.M.: The wavelet transform-domain LMS adaptive filter employing dynamic selection of subband-coefficients. Digital Signal Proc. **69**, 94–105 (2017)

12. Mohammadi, A., Zahiri, S.H.: IIR model identification using a modified inclined planes system optimization algorithm. Artif. Intell. Rev. **48**, 237–259 (2017). https://doi.org/10.1007/s10462-016-9500-z
13. Hartmann, A., Lemos, J.M., Costa, R.S., Vinga, S.: Identifying IIR filter coefficients using particle swarm optimization with application to reconstruction of missing cardiovascular signals. Eng. Appl. Artif. Intell. **34**, 193–198 (2014)
14. Jiang, A.: IIR digital filter design using convex optimization. PhD Dissertation, 2010, Department of Electrical and Computer Engineering, University of Windsor, Canada, Electronic Theses and Dissertations, 432, (2010). https://scholar.uwindsor.ca/etd/432
15. Diniz, P.S.R.: Adaptive Filtering Allgorithms and Practical Implementations. Springer, USA (2008, 2013, 2020, 2020). https://www.springer.com/gp/book/9783030290566
16. Ng, S.C., Leung, S.H., Chung, C.Y., Luk, A., Lau, W.H.: The genetic search approach: a new learning algorithm for adaptive IIR filtering. IEEE Signal Process. Mag. **13**, 38–46 (1996). https://doi.org/10.1109/79.543974
17. Karaboğa, N.: Digital IIR filter design using differential evolution algorithm. EURASIP J. Appl. Signal Proc. **8**(8), 1–9 (2005). https://doi.org/10.1155/ASP.2005.1269
18. Kalinli, A., Karaboğa, N.: A parallel tabu search algorithm for digital filter design. COMPEL Int. J. Comput. Math. Elcctr. Electron. Eng. **24**, 1284–1298 (2005). https://doi.org/10.1108/03321640510615616
19. Karaboğa, N., Çetinkaya, B.: Design of digital FIR filters using differential evolution algorithm. Circuits Syst. Signal Proc. J. **25**, 649–660 (2006). https://doi.org/10.1007/s00034-005-0721-7
20. Krusienski, D.J., Jenkins, W.K.: Design and performance of adaptive systems based on structured stochastic optimization strategies. IEEE Circuits Syst. Mag. **5**(1), 8–20 (2005). https://doi.org/10.1109/MCAS.2005.1405897
21. Krusienski, D.J., Jenkins, W.K.: Adaptive filtering via particle swarm optimization. In: 37th Asilomar Conference on Signals Systems and Computers, pp. 571–575 (2003). https://doi.org/10.1109/ACSSC.2003.1291975
22. Krusienski, D.J., Jenkins, W.K.: Particle swarm optimization for adaptive IIR filter structures. In: Congress on Evolutionary Computation, pp. 965–970 (2004)
23. Kalinli, A., Karaboğa, N.: A new method for adaptive IIR filter design based on tabu search algorithm. Int. J. Electron. Commun. **59**(2), 111–117 (2004). https://doi.org/10.1016/j.aeue.2004.11.003
24. Chen, S., Luk, B.L.: Adaptive simulated annealing for optimization in signal processing applications. Signal Process. ISSN: 0165-1684 **79**, 117–128 (1999)
25. Nelder, J.A., Mead, R.: A simplex method for function minimization. Comput. J. **7**, 308–313 (1965)
26. Shynk, J.J.: Adaptive IIR filtering. In: IEEE ASSP Magazine, pp. 4–21 (1989). https://doi.org/10.1109/53.29644
27. Ljung, L., Söderström, T.: Theory and Practice of Recursive Identification. MIT Press, Cambridge (1983)

Event-Based Looming Objects Detection

Behnam Kamranian and Howard Cheng(✉)

Department of Mathematics and Computer Science, University of Lethbridge, Lethbridge, Alberta, Canada
howard.cheng@uleth.ca

Abstract. An event-based looming objects detection algorithm for asynchronous event-based cameras is presented. The algorithm is fast and accurate both in the detecting the correct number of objects as well as whether the objects are looming.

1 Introduction

In perception, looming is an optical phenomenon in which the size of a given object rapidly expands [18]. Looming often occurs when an object moves closer to the viewer. Fast and accurate looming detection is essential both in nature and robotics.

There are many looming object detection algorithms developed for conventional frame-based cameras (for example, [4,5,7,14,15,20]). However, conventional cameras are limited by their frame rates and also produce redundant data for parts of the scene that remain static. The Dynamic Vision Sensor (DVS) is designed based on the human retina [12]. The DVS is an event-based camera which asynchronously transmits events only when significant changes in the log-luminance of individual pixels are detected. There is no "frame" to collect events over a time interval, and no data is transmitted when there is no significant change. These cameras can react to events faster while their electrical and computational power requirement is lower.

The goal of this paper is to provide a real-time and automatic solution for the problem of detecting multiple looming objects. Optical flow is first computed using an event-based algorithm [17]. Clustering techniques are adapted for asynchronous optical flow events to identify potential objects, and the optical events for each objects are analyzed to determine if the object is looming. Our clustering algorithm does not require *a priori* assumptions on the shapes and number of objects and can adapt to changing number of moving objects in the scene. In addition, our algorithm can run on modest hardware without parallel processing.

2 Preliminaries

2.1 Event-Based Cameras

The Dynamic Vision Sensor (DVS) is a neuromorphic camera which behaves similar to the human visual system by modeling the human retina [12]. Unlike

G. Rozinaj and R. Vargic (Eds.): IWSSIP 2021, CCIS 1527, pp. 82–95, 2022.
https://doi.org/10.1007/978-3-030-96878-6_8

frame-based cameras which collect frames and transmit them synchronously at a fixed frame rate, the DVS asynchronously transmits events as soon as each event occurs. When there are no changes in the log-luminance in the scene, the DVS produces no output. When there is a significant change in the log-luminance of any pixel, the DVS asynchronously reports an event which is described by the coordinates, the timestamp, and the polarity (+/−) of the change. The magnitude of the change is not reported. In the DVS, each pixel can adapt to its own intensity because they are independent from the other pixel sensors. As a result, the DVS has a very high dynamic range. A good survey on the event-based cameras and their applications can be found in [6].

2.2 Event-Based Optical Flow

Ridwan and Cheng [17] presented an event-based optical flow algorithm that detect movements by identifying correlations among events. When objects move in a scene, the log-luminance changes occur mostly at the object boundaries. The pixels of a boundary edge will produce the same polarity along the direction of the motion over a period of time. Therefore, finding events of the same polarity in close proximity in time and space might be an indication of the motion.

The output of this algorithm is a stream of events containing the time, location and direction in eight compass directions ($\vec{v}_0, \ldots, \vec{v}_7$). For each DVS event that arrives, the algorithm searches the eight neighbours of the location for a matching recent event with the same polarity. Experimental results showed that each event can be processed in around two microseconds with very modest hardware. More details of the algorithm can be found in [17].

2.3 Event-Based Object Clustering and Tracking

Barranco et al. presented a method [1] based on mean shift clustering and adapted this algorithm to process asynchronous events. To reduce the required computation time, this method processes events in parallel and in small packets of a few hundred events at a time. Using Kalman filters, this method can track multiple targets [11]. High temporal resolution results in accurate velocity measurements. Despite the advantages of using this method, it does require parallel processing to be feasible for real-time applications.

3 Single Object Looming Detection

In this section, we assume that there is only a single moving object in the scene. We describe an algorithm to detect if this object is looming, and present experimental results demonstrating its effectiveness. The input to our algorithm is the optical flow event stream from the event-based optical flow algorithm [17]. Object boundaries are identified by grouping similar optical flow events—if their angles differ by 45 degrees or less.. The boundary obtained consists mostly of the leading and trailing edge of the moving object. Object movement is classified into three types: moving towards the viewer (looming); moving away from the viewer; or moving sideways. The boundary of a looming object moves away from the center of the object. When an object moves away from the viewer, its boundary moves towards the center. When an object moves sideways, the leading edge of the object moves away from the center while the trailing edge moves towards the center. The arithmetic mean of the locations of all boundary optical flow events is computed to obtain an interior point. Let $\vec{v}$ be the vector associated with the direction in the reported optical flow event, and $\vec{u}$ be the vector from the interior point to the event on the boundary. If $\vec{u} \cdot \vec{v} > 0$, we conclude that $\vec{v}$ is pointing away from the interior (Fig. 1).

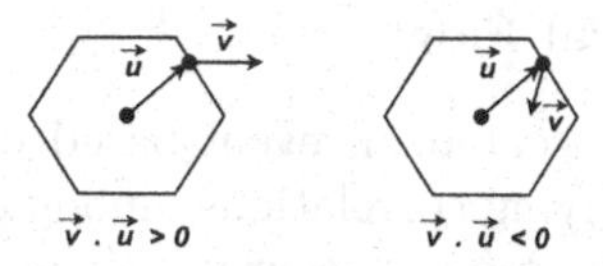

Fig. 1. Determining if an optical flow event is pointing away from the interior point or vise versa using dot product.

Optical flow events are collected into "pseudo-frames" of a certain length, and if the number of events pointing away from the interior is more than twice the number of events pointing towards the interior, our algorithm reports that a looming object is detected in the scene. To reduce the effect of noise, looming should only be reported if there is a significant number of vectors pointing away from the interior compared to the number of pixels in the scene. In our experiments, looming is reported only if the number of vectors pointing away from the interior is at least 0.5% of the total number of pixels in the scene.

The single object looming detection algorithm is shown in Algorithm 1. The algorithm produces an output event only if there is a looming object detected. Otherwise no output is produced. A queue $Q_{(x,y)}$ is used at each pixel to store recent optical flow events. The thresholds required are the length of the pseudo-frame L, and time thresholds T_{low} and T such that only those events between T_{low} and T seconds before the current event are considered recent. A set S is used to collect optical flow events into a pseudo-frame and a global variable $last$ is used to record the timestamp of the last pseudo-frame. On average the complexity is constant for each event.

3.1 Experimental Results

We have performed experiments on our looming object detection algorithm on various scenarios. We have used simple objects with the DVS for these experiments. For visualization, an optical flow event is shown as a blue line moving towards a red dot. A data set was created to test the effectiveness of the algorithm by using thd DVS to capture motion in a scene. A brief of the description of the data set used is given below.

Algorithm 1. Looming detection algorithm.

```
procedure LOOMING(x, y, t, v⃗)
    Remove all events (x', y', t', v⃗') such that t − t' > T from the front of Q_(x,y).
    Add (t, v⃗) to the back of Q_(x,y)
    boundary ← false
    for each (x', y') an 8-neighbour of (x, y) do
        Search in Q_(x',y') for an event e' = (x', y', t', v⃗') such that T_low < t − t' ≤ T
        AND v⃗ and v⃗' are similar
        if NOT found then
            boundary ← true
        end if
    end for
    if NOT boundary then
        return
    end if
    S ← S ∪ {(x, y, v⃗)}
    if t − last > L then
        (c_x, c_y) ← centroid of all events in S
        pos, neg ← 0, 0
        for each (x, y, v⃗) ∈ S do
            d ← ((x, y) − (c_x, c_y)) · v⃗
            if d > 0 then
                pos ← pos + 1
            else if d < 0 then
                neg ← neg + 1
            end if
        end for
        if pos > 2 × neg AND pos > 0.005 × total pixels then
            Report LOOMING at time t
        end if
        S ← {}
        last ← t
    end if
end procedure
```

Round looming object: a round object is moving towards the viewer (Fig. 2(a)).

Round object moving sideways: a round object is moving from right to left (Fig. 2(b)).

Square looming object: a square object is moving towards the viewer (Fig. 2(c)).

Square object moving sideways: a square object is moving from left to right (Fig. 2(d)).

For these experiments, the constants and the thresholds we have chosen experimentally are shown in Table 1. The results of our experiments are shown in Table 2.

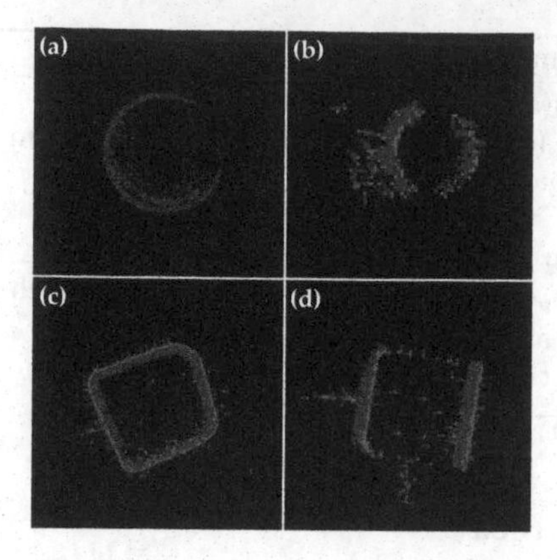

Fig. 2. (a) A round looming object. (b) A round object is moving sideways. (c) A square looming object. (d) A square object is moving sideways.

Table 1. Values of all thresholds and constants for all experiments.

Name	Value	Unit
COLS	180	Pixels
ROWS	190	Pixels
Timestamp threshold (T)	25,000	µs
Low timestamp threshold (T_{low})	100	µs
Length of pseudo-frame (L)	25,000	µs

4 Multiple Object Looming Detection

When there are multiple objects in the same scene, our looming detection algorithm (Sect. 3) fails to detect the looming objects. Our goal in this section is to separate the events in the scene into multiple objects using clustering, and then apply our single object looming detection algorithm to each segmented object.

Clustering algorithms generally require a set of data points to group them into different clusters. However, the optical flow event stream is asynchronous. New events can arrive at any time and old events also need to be removed. Some algorithms solve this problem by using pseudo-frames. Although there are some real-time clustering algorithms adopted for event-based data points [1,13], they required parallel processors or special FPGA hardware for real-time performance. Only the coordinates of the optical flow events are used for clustering. The label of each point along with the centroid of each cluster is reported by the clustering algorithm. We also use L as a parameter for the algorithm to adjust the length of pseudo-frame.

Table 2. Results of single looming object detection.

	Round looming object	Round object moving sideways	Square looming object	Square object moving sideways
Number of events	668530	672201	850175	792649
Video length	3.92 s	2.81 s	5.10 s	10.07 s
Run time/event	1.69 µs	2.82 µs	2.27 µs	2.59 µs
Decision	Looming	Not looming	Looming	Not looming

4.1 *K*-Means Event Clustering

The K-means algorithm [9] is a well-known clustering algorithm. As the number of clusters is not known in advance, the K-means algorithm is executed with different values of $K = 1, \ldots, M$, where M is the maximum number of clusters to consider. A "compactness" measure C is used to compare different outputs produced by clustering algorithms with different values of K:

$$C = \sum_{i=1}^{n} \|x_i - c_{l(x_i)}\|^2, \tag{1}$$

where $l(x_i)$ is the label assigned to event x_i, $\|x_i - c_{l(x_i)}\|$ is the distance between each data point x_i and each cluster's centroid $c_{l(x_i)}$. By plotting the compactness as a function of K, we can find the "elbow point" where the rate of reduction changes drastically [8,10,21]. The elbow point is a good candidate for the number of clusters. Figure 3 shows the elbow point and the change in the compactness measure as the number of clusters increases. There is no consensus on a mathematically rigorous definition of the elbow point [10]. The Kneedle algorithm [19] had been proposed for finding the elbow point, but experiments show that it was not well-suited for the data arising in our application.

4.2 The Elbow Method

Heuristically, the elbow point is the point at which the angle of the curve is the greatest (Fig. 3). Only those points in which the decrease in compactness is

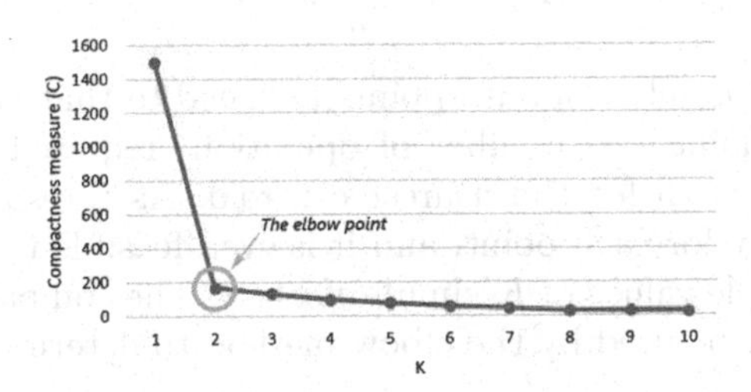

Fig. 3. The elbow point method.

greater than the average decrease over all values of K are considered, and the point with the largest angle is reported as the number of clusters (Algorithm 2). In the algorithm, M is the maximum number of clusters to consider, and C_i is the compactness measure when the K-means algorithm is used to cluster the events into i clusters.

Algorithm 2. The elbow method

procedure ELBOW($M, C_1, \ldots, C_M$)
  $avg = \frac{C_1 - C_M}{M-1}$
  $\beta_{max} = 0$
  for $i = 1, \ldots, M-1$ **do**
    $\Delta_i = C_{i+1} - C_i$ ▷ Note: $\Delta_i < 0$
  end for
  for $i = 1, \ldots, M-2$ **do**
    if $\Delta_i \le \Delta_{i+1}$ AND $-\Delta_i \ge avg$ **then**
      $\beta = \arccos\left(\frac{(1,\Delta_i)\cdot(1,\Delta_{i+1})}{\|(1,\Delta_i)\|\cdot\|(1,\Delta_{i+1})\|}\right)$
      if $\beta > \beta_{max}$ **then**
        $\beta_{max} = \beta$
        $K = i + 1$
      end if
    end if
  end for
  return K
end procedure

4.3 Sequential *K*-Means Clustering

Sequential K-means clustering is a variation of the standard K-means clustering algorithm that processes one data point at a time and update the clusters' centroids at each step [3]. centroids at a particular time. When a new data point x is received, the algorithm chooses the centroid c_i closest to x and adds x to the corresponding cluster. The centroid c_i is updated by

$$c_{i+1} = c_i + \frac{1}{n}(x - c_i), \tag{2}$$

where n is the total number of data points assigned to that cluster, including x.

For each data point, the number of operations required is proportional to K because of the search for the nearest centroid. As a result, the update can be done very quickly for each point, and it is even feasible to perform K-means clustering for multiple values of K simultaneously. The compactness measure for each value of K can be used by the elbow method to determine the appropriate number of clusters.

4.4 Cluster Merging

When objects are too large, the clustering algorithms may fail to detect the correct number of clusters by dividing them into separate clusters. This is because these algorithms try to minimize the average squared distance between each data point and the centroid of the clusters.

As a solution to this problem, we can merge these clusters to form a single cluster. Clusters that are connected as 8-neighbours are merged into one connected component as a new cluster.

5 Experiments and Results

Tje different proposed clustering algorithms described in Sect. 4 are evaluated. The algorithms are tested with event streams generated from both captured and simulated scenarios. The algorithms are tested with data sets shown in Table 3. The captured event streams were obtained with the DVS specified in Table 4.

Table 3. Captured data sets.

Data set	Description	Number of polarity events	Number of optical flow events	Number of pseudo-frames
1	A single ball is falling	14900	9074	18
2	Two round objects are moving sideways	113240	81214	291
3	A round object is looming	79850	52786	73
4	Two balls are rolling sideways	22026	14831	18
5	Four round objects are looming	40900	26302	35

All algorithms were implemented in C++ and the OpenCV library [2]. To evaluate the correctness of the cluster detection algorithms we reported the number of pseudo-frames in which the correct number of clusters was detected. To evaluate the quality of each detected cluster, we manually checked the labelling in each pseudo-frame. We manually labelled the looming results of each pseudo-frame and we compared them with the results generated by the algorithms. A looming object detected correctly is a true positive, while a true negative occurs when the algorithm produce no output when there are no-looming objects. The commonly used Recall and Precision measures [16] are computed.

Table 4. The DVS specifications used for experiments.

Name	Value
Model	DVS 240 B
I/O	USB2.0
Power consumption	Low/high activity: 30/60 mA @ 5 VDC
Number of columns (*COLS*)	180 pixels
Number of rows (*ROWS*)	190 pixels

The figures show the detected clusters in different colours and depict the centroid of each cluster by a dot surrounded by a circle with the same colour of its cluster. The detected looming clusters are shown as yellow circles on their centroids (Fig. 4).

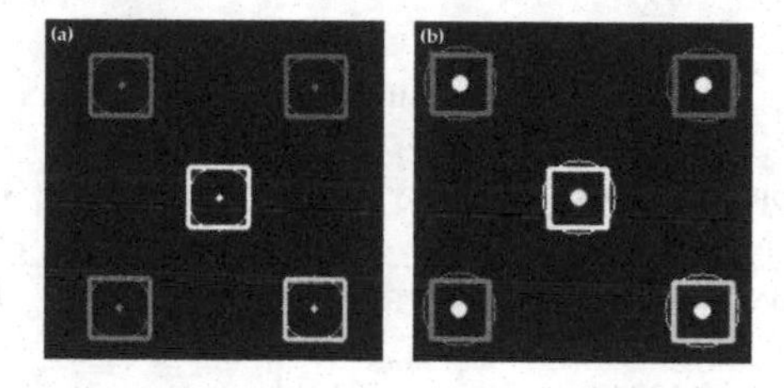

Fig. 4. (a) five detected clusters. (b) Five detected looming clusters

In Data Set 1, a ball is falling in front of the camera. The goal of this experiment is to determine whether the clustering algorithms is capable of detecting a single cluster. Figure 5 shows the optical flow events and output of clustering for one of the pseudo-frames of this data set. All of the results of experiments on captured data sets are reported in Table 5. Both sequential and OpenCV's K-means algorithms failed to detect the correct number of clusters in all pseudo-frames. Overall applying the merge algorithm enhanced the results drastically. Using the elbow method and sequential K-means algorithm, only five pseudo-frames have an the incorrect number of detected clusters. In this data set the sequential K-means is at least 54 times faster than OpenCV's K-means and 10 times faster than the mean shift algorithm without parallel processing (Sect. 2.3).

In Data Set 2, two round objects are moving sideways. The goal of this experiment is to compare the accuracy of algorithms when objects move straight versus when objects are rolling (Data set 6) in front of the camera. Figure 6 shows the optical flow events and clustering output for a pseudo-frame of this data set. All algorithms were able to detect the correct number of clusters in all pseudo-frames. The fastest algorithm is sequential K-means which on average took about 0.6 ms to process each pseudo-frame.

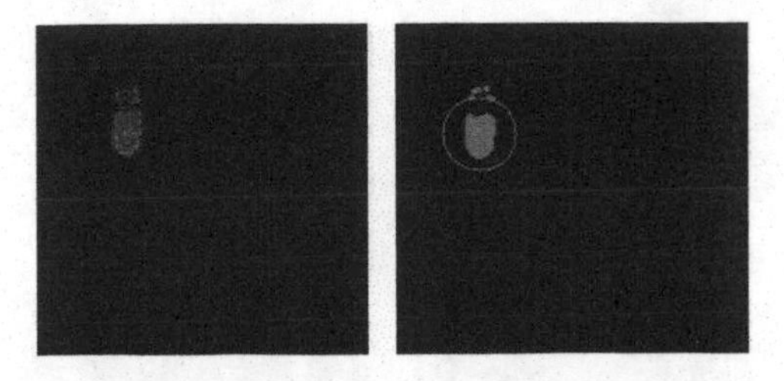

Fig. 5. A single ball is falling.

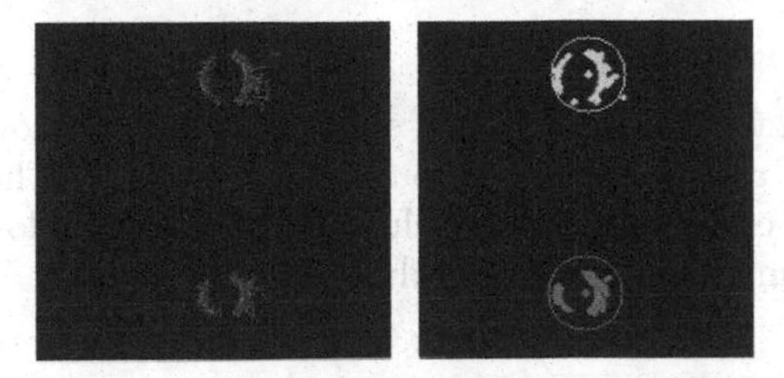

Fig. 6. Two round objects are moving sideways.

In Data Set 3, a ball is approaching the camera. We moved the ball very close to the camera to see how the algorithms can perform when the dimensions of the objects are large or when they are close to the camera. None of the algorithm was able to detect the correct number of clusters. The reason is that when the object is so close to the camera, the algorithms separate it to multiple clusters to decrease the average squared distance from each event to the computed centroid. Figure 7 shows a pseudo-frame in this situation. Applying the merge algorithm enhanced the results. Figure 8 shows a pseudo-frame in which the merge algorithm was able to merge multiple clusters to a single cluster. However, due to both noises and lack of events in some parts of the object, the clustering algorithms were not able to detect a single cluster even by using the merge algorithm. The sequential K-means algorithm is again the fastest method and processed each pseudo-frame about 60 times faster than OpenCV's K-means algorithm.

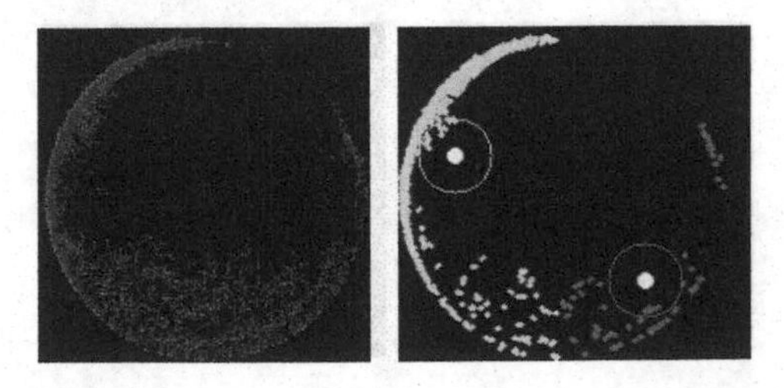

Fig. 7. A single looming ball which is incorrectly detected as two clusters. The merge algorithm was not applied.

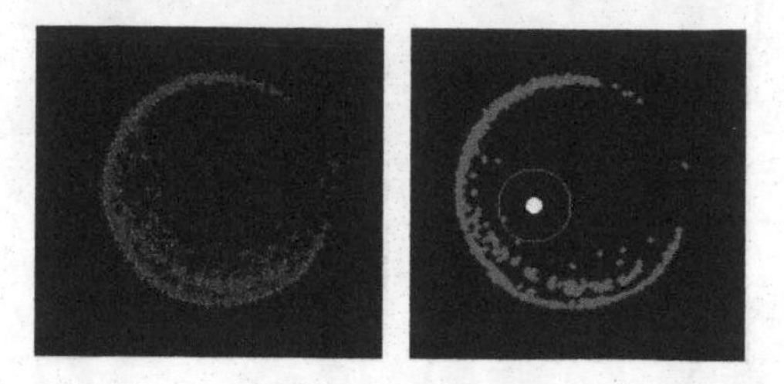

Fig. 8. Single ball is looming and detected as a single cluster by applying the merge algorithm.

For Data Set 4, two balls are rolling sideways. Figure 9 shows a pseudo-frame of the optical flow and clustering output of this data set. The algorithms were able to detect the correct number of clusters in most pseudo-frames. Applying our merge algorithm enhances the results further.

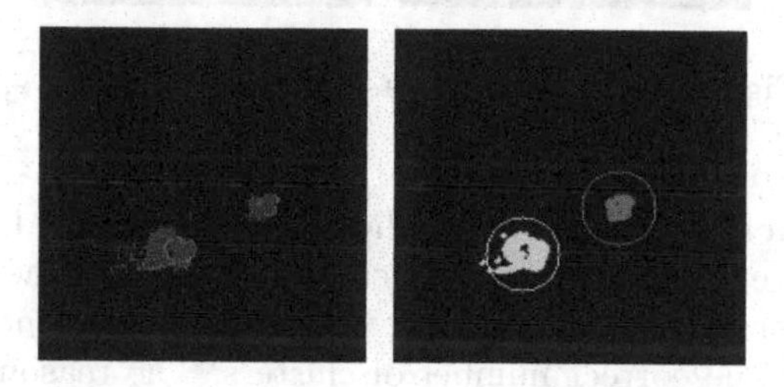

Fig. 9. Two balls are rolling sideways.

For Data Set 5, the camera is moving toward four round objects in a solid white background. Figure 10 shows the optical flow events and clustering output for a single pseudo-frame of this case. Our algorithm was able to detect the correct number of clusters, though the recall rate is low because the camera is approaching the objects from an angle and it classifies the motion as sideways instead of looming.

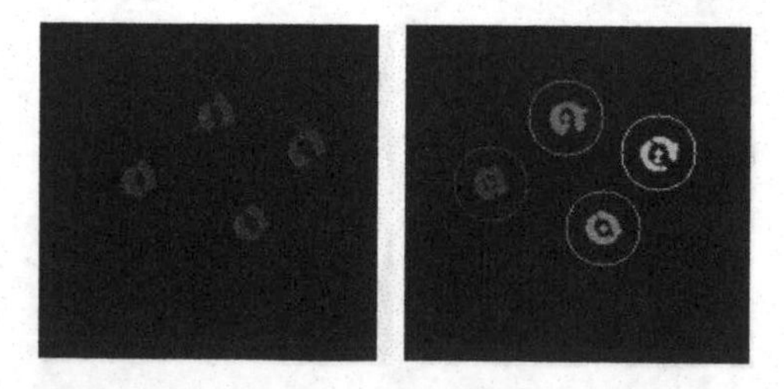

Fig. 10. Four round objects are looming.

Table 5. The results of experiments on captured data sets.

Dataset	Clustering methods	Correct number of clusters (%)	Looming correctness (%)		Processing time/Pseudo-frame (*ms*)
			Precision	Recall	
1	*K*-means	88.88	72.22	100.00	28.125
	Seq. *K*-means	72.22	66.66	100.00	0.543
2	*K*-means	100.00	97.93	100.00	26.944
	Seq. *K*-means	100.00	96.90	100.00	0.693
3	*K*-means	56.16	95.71	91.17	41.596
	Seq. *K*-means	41.09	98.59	95.89	0.799
4	*K*-means	100.00	88.88	100.00	34.514
	Seq. *K*-means	100.00	80.55	100.00	0.632
5	*K*-means	25.71	100.00	7.14	36.391
	Seq. *K*-means	100.00	100.00	5.71	0.667

The results of the experiments indicate the advantages of using sequential *K*-means algorithm compared to other methods. This is a real-time algorithm, and does not require any parameter adjustment. It can automatically adapt itself in all experiments cases to changing number of objects and movement types. Compared to other clustering algorithms, it is much faster and can process each pseudo-frame in less than 0.8 ms depending on the size of the input. This is at least 30 times faster than OpenCV's *K*-means algorithm. In addition, the sequential *K*-means algorithm achieved the highest accuracy in cluster detection compared to other algorithms in most data sets.

6 Conclusion

We presented a real-time looming object detection algorithm using event-based camera. It does not require any a priori knowledge of the number of objects in the scene and can adapt to changing number of objects. The proposed algorithm is significantly faster than the conventional *K*-means algorithm, and the accuracy for looming object detection is similar.

While our algorithm performs very well when objects are looming directly towards the camera, recall rate is low when the objects are looming towards the camera at an angle. Future works will address this limitation.

References

1. Barranco, F., Fermüller, C., Ros, E.: Real-time clustering and multi-target tracking using event-based sensors. CoRR abs/1807.02851 (2018). http://arxiv.org/abs/1807.02851
2. Bradski, G.: The OpenCV library. Dr. Dobb's J. Softw. Tools (2000). https://github.com/opencv/opencv/wiki/CiteOpenCV
3. Duda, R., Hart, P., Stork, D.: Pattern classification, 2nd edn. Wiley-Interscience (2000)
4. Fülöp, T., Zarándy, A.: Bio-inspired looming object detector algorithm on the eye-RIS focal plane-processor system. In: 2010 12th International Workshop on Cellular Nanoscale Networks and Their Applications (CNNA), pp. 1–5. IEEE (2010)
5. Fülöp, T., Zarándy, A.: Bio-inspired looming direction detection method. In: 2012 13th International Workshop on Cellular Nanoscale Networks and Their Applications (CNNA), pp. 1–6. IEEE (2012)
6. Gallego, G., et al.: Event-based vision: a survey. IEEE Trans. Pattern Anal. Mach. Intell. 1 (2020). https://doi.org/10.1109/TPAMI.2020.3008413
7. Gil-Jiménez, P., Gómez-Moreno, H., López-Sastre, R., Bermejillo-Martín-Romo, A.: Estimating the focus of expansion in a video sequence using the trajectories of interest points. Image Vis. Comput. **50**, 14–26 (2016)
8. Goutte, C., Toft, P., Rostrup, E., Nielsen, F., Hansen, L.: On clustering FMRI time series. NeuroImage **9**(3), 298–310 (1999). https://doi.org/10.1006/nimg.1998.0391. http://www.sciencedirect.com/science/article/pii/S1053811998903913
9. Hartigan, J., Wong, M.: Algorithm as 136: a k-means clustering algorithm. J. R. Stat. Soc. Ser. C (Appl. Stat.) **28**(1), 100–108 (1979)
10. Ketchen, D., Shook, C.: The application of cluster analysis in strategic management research: an analysis and critique. Strateg. Manag. J. **17**(6), 441–458 (1996)
11. Li, X., Wang, K., Wang, W., Li, Y.: A multiple object tracking method using Kalman filter. In: The 2010 IEEE International Conference on Information and Automation, pp. 1862–1866, June 2010. https://doi.org/10.1109/ICINFA.2010.5512258
12. Lichtsteiner, P., Posch, C., Delbruck, T.: A 128 $\times$ 128 120 *db* 15 μs latency asynchronous temporal contrast vision sensor. IEEE J. Solid-State Circuits **43**(2), 566–576 (2008)
13. Linares-Barranco, A., Gómez-Rodríguez, F., Villanueva, V., Longinotti, L., Delbrück, T.: A USB3.0 FPGA event-based filtering and tracking framework for dynamic vision sensors. In: 2015 IEEE International Symposium on Circuits and Systems (ISCAS), pp. 2417–2420, May 2015. https://doi.org/10.1109/ISCAS.2015.7169172
14. Pantilie, C., Nedevschi, S.: Real-time obstacle detection in complex scenarios using dense stereo vision and optical flow. In: 13th International IEEE Conference on Intelligent Transportation Systems, pp. 439–444. IEEE (2010)
15. Park, S.S., Sowmya, A.: Autonomous robot navigation by active visual motion analysis and understanding. In: Proceedings of IAPR Workshop on Machine Vision Applications (1998)
16. Powers, D.: Evaluation: from precision, recall and F-measure to ROC, informedness, markedness and correlation (2011)
17. Ridwan, I., Cheng, H.: An event-based optical flow algorithm for dynamic vision sensors. In: Karray, F., Campilho, A., Cheriet, F. (eds.) ICIAR 2017. LNCS, vol. 10317, pp. 182–189. Springer, Cham (2017). https://doi.org/10.1007/978-3-319-59876-5_21

18. Rind, F., Simmons, P.: Seeing what is coming: building collision-sensitive neurones. Trends Neurosci. **22**(5), 215–220 (1999)
19. Satopaa, V., Albrecht, J., Irwin, D., Raghavan, B.: Finding a "kneedle" in a haystack: detecting knee points in system behavior. In: 2011 31st International Conference on Distributed Computing Systems Workshops, pp. 166–171, June 2011. https://doi.org/10.1109/ICDCSW.2011.20
20. Subbarao, M.: Bounds on time-to-collision and rotational component from first-order derivatives of image flow. Comput. Vis. Graph. Image Process. **50**(3), 329–341 (1990)
21. Thorndike, R.: Who belongs in the family? Psychometrika **18**(4), 267–276 (1953). https://doi.org/10.1007/BF02289263

Moment Transform-Based Compressive Sensing in Image Processing

Theofanis Kalampokas and George A. Papakostas(✉)

MLV Research Group, Department of Computer Science, International Hellenic University, Kavala, Greece

{theokala,gpapak}@cs.ihu.gr

Abstract. Over the last decades, images have become an important source of information in many domains, thus their high quality has become necessary to acquire better information. One of the important issues that arise is image denoising, which means recovering a signal from inaccurately and/or partially measured samples. This interpretation is highly correlated to the compressive sensing theory, which is a revolutionary technology and implies that if a signal is sparse then the original signal can be obtained from a few measured values, which are much less, than the ones suggested by other used theories like Shannon's sampling theories. A strong factor in Compressive Sensing (CS) theory to achieve the sparsest solution and the noise removal from the corrupted image is the selection of the basis dictionary. In this paper, Discrete Cosine Transform (DCT) and moment transform (Tchebichef, Krawtchouk) are compared in order to achieve image denoising of Gaussian additive white noise based on compressive sensing and sparse approximation theory. The experimental results revealed that the basis dictionaries constructed by the moment transform perform competitively to the traditional DCT. The latter transform shows a higher PSNR of 30.82 dB and the same 0.91 SSIM value as the Tchebichef transform. Moreover, from the sparsity point of view, Krawtchouk moments provide approximately 20–30% more sparse results than DCT.

Keywords: Compressive sensing · Image moments · Denoising · Tchebichef moments · Krawtchouk moments · DCT

1 Introduction

Compressive Sensing is an important contribution for the reason that it overthrows the traditional sampling theory of Shannon and Nyquist where in order to reconstruct a signal without error the sampling rate should be at least twice the signal's maximum frequency component. Compressive sensing as introduced by Donoho et al. [1], Candes et al. [2] gave the potential to reconstruct a full signal even if it is sparse. With these results over the past decades has been done a significant effort in the development of techniques in many domains around CS and sparse representation theory because it affects many domains from the perspective of storage to the recovery of corrupted information despite the structure of the signal. With the introduction of this theory many image processing

G. Rozinaj and R. Vargic (Eds.): IWSSIP 2021, CCIS 1527, pp. 96–107, 2022.
https://doi.org/10.1007/978-3-030-96878-6_9

and computer vision problems are affected not only in the compression but to denoising or inpainting of images. Besides CS-based denoising or inpainting, there is a variety of research work that has been done in image watermarking [3], image hiding [4].

In general CS and sparse representation theory makes two conjectures: that any natural image can be sparse in a basis or a dictionary, constructed by Fourier, wavelet, DCT, or any other transforms, where only a few transform coefficients are significant and the rest are zero or negligible, and that the measurement basis is incoherent with the basis, which the image is sparse [5].

In image denoising, the techniques can be divided into two groups: *spatial-domain* and *transform-domain* methods. In the former category are included methods based on the Perona-Malik equation, which brings good noise reduction and edges preservation [6], or the bilateral filter technique, which acquires the original image with noise reduction [7]. The transform-domain denoising methods include some modern algorithms arising from CS and sparse representation theory where the sparsity of the signal is exploited through some transform. For example, Strack et al. [8] proposed a new geometric multiscale transform Curvelet and in comparison with wavelet, it achieved better results in image denoising.

There is a huge variety of implementations that have been proposed around the above theory, from the different transformations in the basis dictionary to the creations of new algorithms that are applied to several data structures for different domains. In this paper, the moment transform is proposed in CS and sparse representation theory for the description of images in a denoising task, in comparison with classic transformation Discrete Cosine Transform (DCT). More specifically, two different moment families are used the Tchebichef and Krawtchouk moments. The former moment family is known for its noise invariance and later one for its capabilities to describe a signal locally.

The main contribution of this paper is the proposal for the first time, the moment transform in the denoising problem based on CS and sparse representation theory, towards advantaging from the noise tolerance and local description of the examined moment families. From the experimental results, it is proven that moment transform brings more sparse results and thus more compressible than DCT according to the increase of noise level, although brings similar reconstruction quality.

The rest of this paper is organized as follows: Sect. 2 presents the theory of transform-based compressive sensing. Section 3 describes the moment transform, with emphasis on the Tchebishef and Krawtchouk moments used in this work, along with the proposed methodology. The proposed moment-based compressive sensing methodology is described in Sect. 4 and experimental results are discussed in Sect. 5. Finally, Sect. 6 concludes this work.

2 Transform-Based Compressive Sensing

The sampling technique in CS theory gives the potential to sample a signal at rates proportional to the amount of information in the signal by exploiting the sparsity properties of signals. Considering that a signal e.g. an image or can have a sparse representation when expressed w.r.t. some basis or a dictionary $\boldsymbol{\Psi} \in \mathbb{R}^{n \times m}$. This dictionary shows a complete structure where $n = m$, or an overcomplete structure where $n < m$. Then a signal $\boldsymbol{x}$ can be represented as a linear combination of basis $\boldsymbol{\Psi}$ atoms, with $\boldsymbol{a} \in \mathbb{R}^{n\times m}$ being the projection coefficients of signal $\boldsymbol{x}$ in $\boldsymbol{\Psi}$ domain, expressed in as:

$$\boldsymbol{x} = \boldsymbol{\Psi a}, \tag{1}$$

then the compressed signal $\boldsymbol{y}$ is derived as follows:

$$\boldsymbol{y} = \boldsymbol{\Phi x} = \boldsymbol{\Phi \Psi a} = \boldsymbol{Da}. \tag{2}$$

If the processed signal is contaminated with noise or it is corrupted with irrelevant content (2) can be expressed as:

$$\boldsymbol{y} = \boldsymbol{Da} + \boldsymbol{z}. \tag{3}$$

In (3) $\boldsymbol{z}$ represents noise measurements of any type and $\boldsymbol{D}$ is the sensing matrix. The meaning of compression is that signal $\boldsymbol{y}$ will be smaller than the original signal x. Thus the task is to recover $\boldsymbol{x}$ based on $\boldsymbol{y}$, $\boldsymbol{D}$, and $\boldsymbol{\Psi}$. The sensing matrix $\boldsymbol{D}$ guarantees the signal sparsity or the separation between noise and signal content and must satisfy the Restricted Isometry Property (RIP) for any k-sparse signal $\boldsymbol{x}$ as proposed by [9] and expressed as:

$$(1-\delta)\|\boldsymbol{a}\|^2 \leq \|\boldsymbol{Da}\|^2 \leq (1+\delta)\|\boldsymbol{a}\|^2, \tag{4}$$

where $\delta_k \in (0, 1)$ is the Restricted Isometry Constant (RIC).

The above theory highlights the importance of the transformation selection that will describe the image in a more sparse form. The intrinsic properties of the signals may not be suitably interpreted by all transformations, which makes sparsity not feasible. The transformation selection as the basis dictionary of the signal for different tasks is a topic that still concerns the research community. Ansari et al. [10] implemented a comparison between various transformations e.g. Wavelet, Curvelet, and Contourlet for denoising remote-sensed images contaminated with Gaussian noise. They concluded that Curvelet denoising preserves better the sharpness of the boundaries. Starck et al. [11] proposed Undecimated Wavelet Transform (UWT) where it is proven that overcomes the disadvantage of the discrete wavelet transform regarding its shift invariance property. Wang et al. [12] proposed Shearlet Transform (ST) in CS theory, which is a directional multiresolution transformation, providing higher PSNR in different sampling ratios than the Wavelet Transform (WT). The shifting of the input signal causes small changes to the transform coefficients, which results in a bad representation of edges and borders. Dragotti et al. [13] introduced Directionlets, which is a transformation that provides an efficient interpretation for the nonlinear approximation of images and compared to the WT provides better PSNR with similar complexity.

3 Moment Transform

Image moments are the coefficients of the Moment Transform (MT) that has achieved a significant contribution around signal processing in a variety of domains [14]. Their advantage arises from the robustness in describing an image with fewer coefficients than the actual size of the image, which is a characteristic that matches the CS and sparse representation theory. Among several moment types, the orthogonal moments include orthogonal polynomials as kernel functions, owing desirable properties in both continuous and discrete coordinate spaces [15]. Due to the orthogonality property, image moments provide a more compact representation of an image and robustness to noise.

The first orthogonal moments were expressed in continuous space and it was Zernike, Pseudo-Zernike, Fourier-Mellin, Legendre and are used in many applications for feature description as Kadir et al. [16]. The disadvantage of continuous orthogonal moments is the approximation errors that arise for the reason of the coordinate normalization and space granulation procedures. To overcome this disadvantage discrete orthogonal moments have proposed where defined inside the discrete coordinate system of the image, with the Tchebichef [17], Krawtchouk [18], and dual Hahn [19] moments being the most representative discrete moment families. As a transformation image moments have been proposed in many applications from pattern recognition [20] and adversarial computer vision [21] to the interpretation of EEG signals for seizure classification [22] where the proposed method is proven the robustness of the moment transform in the presence of noise.

Tchebichef moments are robust to high noise levels and object description in comparison with other image orthogonal moments. The Tchebichef Moments (TMs) for an image with *NxN* pixels size are defined as:

$$T_{nm} = \frac{1}{\overline{\rho}(n, N)\tilde{\rho}(m, N)} \sum\nolimits_{x=0}^{N-1} \sum\nolimits_{y=0}^{N-1} \tilde{t}_n(x)\tilde{t}_m(y) f(x, y). \tag{5}$$

In (5) the first term $\overline{\rho}(n, \mathrm{K})$ corresponds to the normalized norm of the Tchebichef polynomials and the term $\tilde{t}_n(x)$ is the normalized Tchebichef polynomials defined as:

$$\tilde{t}_n(x) = t_n(i)/\beta(n, N), \tag{6}$$

with

$$\begin{aligned} t_n(i) &= (1-N)_n 3F_2(-n, -x,\ 1+n;\ 1, 1-N;\ 1 \\ &= n! \sum\nolimits_{k=0}^{n} (-1)^{n-k} \binom{N-1-k}{n-k} \binom{n+k}{n} \binom{x}{k}, \end{aligned} \tag{7}$$

and $\beta(n, N)$ is usually equal to N^n.

Krawtchouk moments [18] are another family of discrete orthogonal moments, characterized by their high local representation capabilities. The locality characteristic of Krawtchouk moments is controlled by the parameters p_1, p_2 and express the spread of the coefficient calculation in an image. The Krawtchouk moments (KMs) of order n and repetition m for a *NxN* pixels are computed with the following formula:

$$K_{nm} = \sum\nolimits_{x=0}^{N-1} \sum\nolimits_{y=0}^{N-1} \overline{K}_n(x; p_1, N-1)\overline{K}_m(y; p_2, N-1) f(x, y), \tag{8}$$

where $\overline{K}_m$ are the weighted Krawtchouk polynomials defined in [18].

4 Compressive Sensing Based on Moment Transform

As discussed above, image moments are robust to noise presence and describe the content of an image in a compact way. Based on the theory, there are two alternative strategies for denoising: one is by *compression*, which is achieved through the transformation where in this occasion is the *moment transform* and the second is the optimal sparse representation of the signal through the transformation. The computed basis dictionary for Tchebichef and Krawtchouk moments, which are examined in this study can be expressed as:

$$\Phi_{nm} = \left[Poly_n\right]^T Poly_{m,}, \tag{9}$$

where $Poly_n$ and $Poly_m$ are the nth and m^{th} order orthogonal polynomials of any moment family, respectively. According to (9) the basis dictionary of a specific moment family is extracted in order to be used in sparse coding or reconstruction algorithm to recover an image.

Next, the reconstruction process takes place through a sparse recovery algorithm from the family of greedy algorithms, which are faster than convex algorithms. These algorithms rely on an iterative approximation of the image coefficients, by obtaining an improved estimation of sparse representation of the image at each iteration that attempts to account for the mismatch to the measured data. The algorithm that is used in this work is the Orthogonal Matching Pursuit (OMP), which is an improvement of the matching pursuit algorithm. It computes the inner product of the residue and the measurement matrix and then selects the index of the maximum correlation column, extracts this column in each iteration and adds it to the selected set of atoms. Then an orthogonal projection is performed over the subspace of previously selected atoms, which provides a new approximation vector used to update the residual. Since in the measurement matrix the columns are orthogonal there will be no column selected twice.

As presented in (2) there is an attempt to select the fewer columns of $\boldsymbol{D}$ that participate in $\boldsymbol{y}$. With OMP as a reconstruction algorithm, there are two steps that need to be fulfilled according to:

$$\lambda_t = \arg \max_{j=1,2...,N} \left|\left\langle \boldsymbol{r}_{t-1}, \boldsymbol{\Psi}_j \right\rangle\right|, \tag{10}$$

where λ_t is the index that solves the above optimization problem, which consists of the terms $\boldsymbol{r}$ and $\boldsymbol{\Psi}$ which are the residual and the transformation basis, respectively. Next, a least squares problem is solved to obtain a new image estimation:

$$x_t = \arg \min_x \|\boldsymbol{D}_t\boldsymbol{x} - \boldsymbol{y}\|_2. \tag{11}$$

Then a new approximation of data and the new residual are calculated:

$$a_t = D_t x_t, \ (13) \quad r_t = y - a_t. \tag{12}$$

The above are calculated for a number (t) of iterations and terminates when a stopping criterion is met.

5 Experiments

In this paper, the denoising of five benchmark images *Lena, Barbara, Baboon, Pirate, and Peppers*, contaminated with Gaussian additive noise, using sparse coding with OMP algorithm is applied, and Tchebichef, Krawtchouk, and DCT basis dictionaries are examined. The experiments are implemented in Python with the usage of Scikit-learn library and executed in a Ryzen 3700 CPU with 16 GB RAM. Figures 1, 2 and 3, depict the basis dictionaries for each transformation.

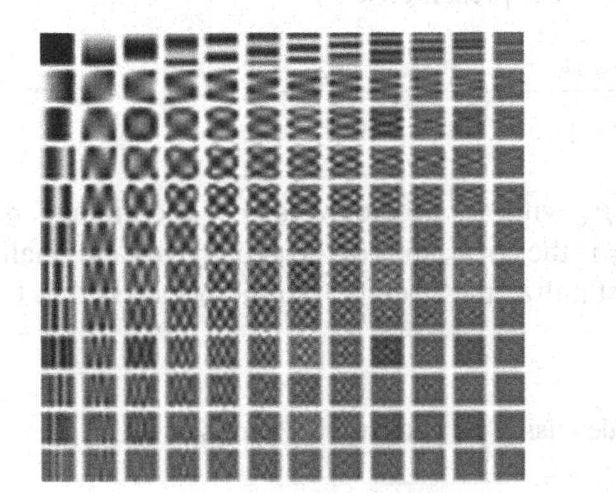

Fig. 1. Tchebichef basis dictionary

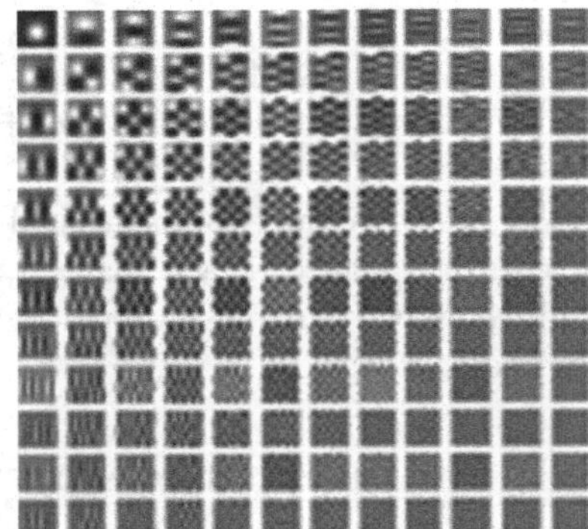

Fig. 2. Krawtchouk basis dictionary

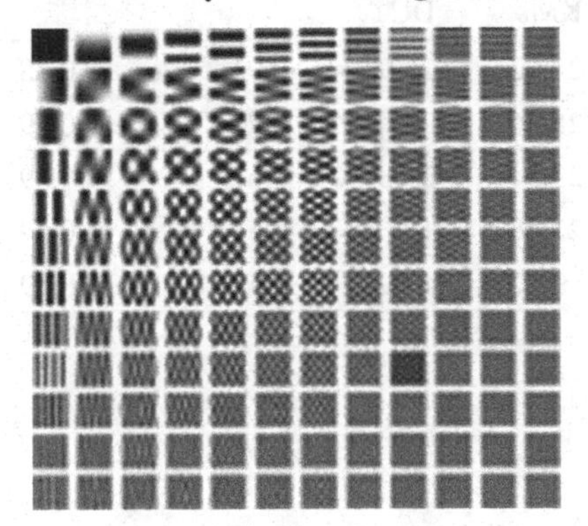

Fig. 3. DCT basis dictionary

Next, the input images are described in overlapped patches of 12×12 pixels size in all image dimensions, in order to achieve a redundant structure between image information and noise presence. The selection of this patch dimension is calculated by the square root of image dimensions, which are 144×144. Such patches are presented as 144-dimensional vectors of gray-scale values. The DC value and the mean of the gray-scale values were subtracted from each vector as a preprocessing step. The result is a linear dependency between the components of the observed data and therefore the dimension of the data was reduced by one. The denoising results for each benchmark image and different noise ratios are presented in the following Table 1, 2, 3, 4 and 5 and Figs. 4,

5, 6, 7 and 8. It is worth noting that the PSNR (in dB) and SSIM indices are used to evaluate the quality of the denoised images. First the mathematical expression of PSNR is:

$$\mathrm{PSNR} = 10\log_{10}\frac{N^2}{\sum_{i=0}^{N-1}\sum_{j=0}^{N-1}\left(x_{i,j}^{(p)} - x_{i,j}^{(0)}\right)^2}. \tag{13}$$

In (13) N is the maximum value of the pixel and the denominator is the reconstruction Mean Squared Error. The definition of SSIM is presented as:

$$\mathrm{SSIM}(x, y) = \frac{(2\mu_x\mu_y + c_1)(2\sigma_{xy} + c_2)}{\left(\mu_x^2 + \mu_y^2 + c_1\right)\left(\sigma_x^2 + \sigma_y^2 + c_2\right)}. \tag{14}$$

In (14) x and y are the two images and μ_x with μ_y are the average of each image, σ_x^2 with σ_y^2 is the variance of each image, σ_{xy} is the covariance of the two images and finally c_1 and c_2 are two variables in order to stabilize the operation in the occasion that the denominator is weak.

Table 1. TMs, KMs, DCT denoising performance in Peppers.

Noise ratio	PSNR-TMs	PSNR-KMs	PSNR-DCT	SSIM- TMs	SSIM- KMs	SSIM- DCT
0.1	27.27	27.02	26.92	0.91	0.88	0.91
0.2	24.35	25.36	23.96	0.78	0.79	0.78
0.3	22.32	23.54	21.81	0.67	0.69	0.66
0.4	20.16	21.48	19.64	0.57	0.58	0.56
0.5	18.21	19.52	17.98	0.48	0.49	0.47

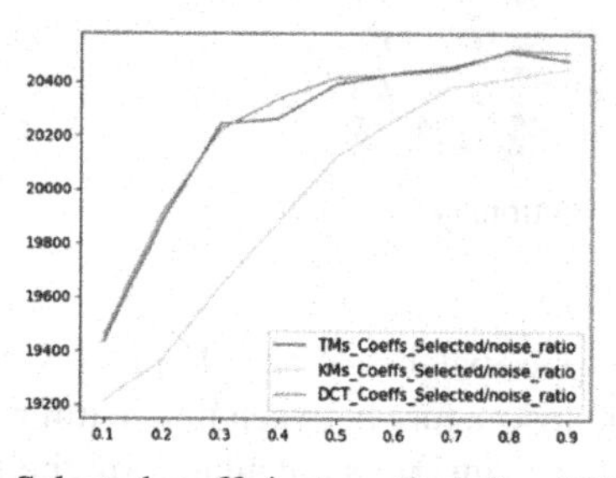

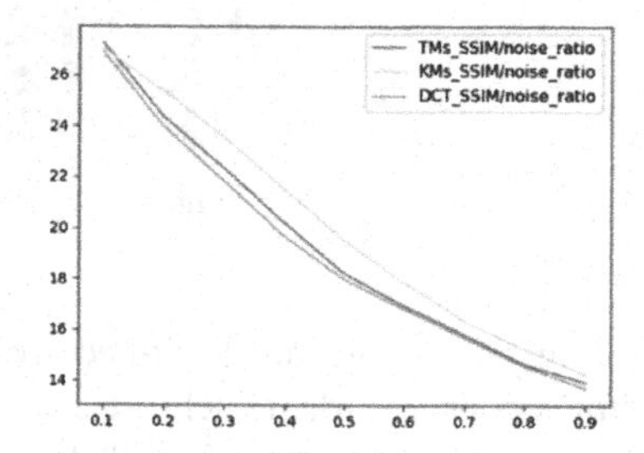

Fig. 4. Selected coefficients on the left and SSIM on the right for TMs, KMs and DCT in Peppers.

From the above results, it is concluded that moment transform in comparison with DCT is bringing more sparse results and almost equal PSNR and SSIM metrics. TMs and DCT perform almost equal for both PSNR and SSIM, in the case of Peppers, Baboon, and Pirate images, but TMs achieve more sparse results than DCT. Next in Barbara and

Table 2. TMs, KMs, DCT denoising performance in Barbara.

Noise ratio	PSNR-TMs	PSNR-KMs	PSNR-DCT	SSIM- TMs	SSIM- KMs	SSIM- DCT
0.1	27.69	26.51	27.96	0.88	0.85	0.88
0.2	26.19	25.47	26.36	0.81	0.79	0.81
0.3	24.26	23.82	24.29	0.72	0.71	0.72
0.4	21.68	21.84	21.71	0.61	0.62	0.62
0.5	20.38	20.13	19.86	0.54	0.53	0.53

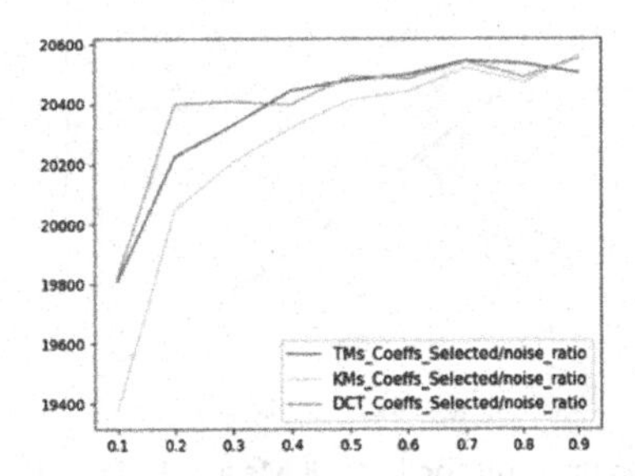

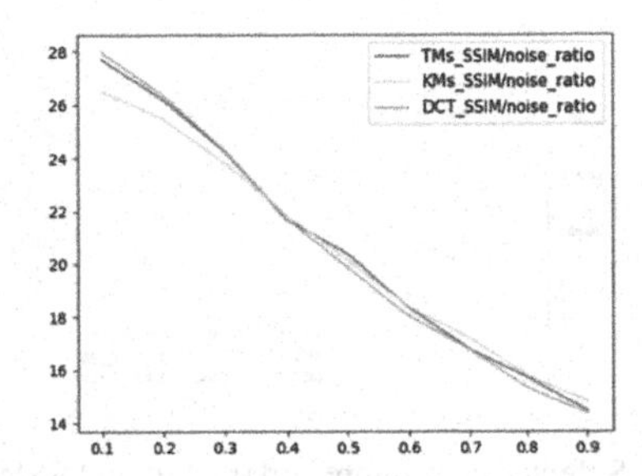

Fig. 5. Selected coefficients on the left and SSIM on the right for TMs, KMs and DCT in Barbara.

Table 3. TMs, KMs, DCT denoising performance in Lena.

Noise ratio	PSNR-TMs	PSNR-KMs	PSNR-DCT	SSIM- TMs	SSIM- KMs	SSIM- DCT
0.1	30.47	29.84	30.82	0.91	0.89	0.91
0.2	28.04	27.38	28.08	0.81	0.79	0.79
0.3	25.22	24.87	25.07	0.68	0.67	0.66
0.4	22.88	22.93	22.89	0.57	0.58	0.57
0.5	20.75	20.99	20.74	0.48	0.49	0.48

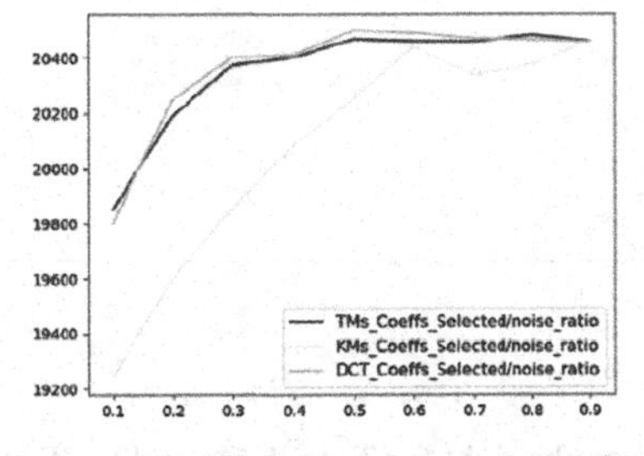

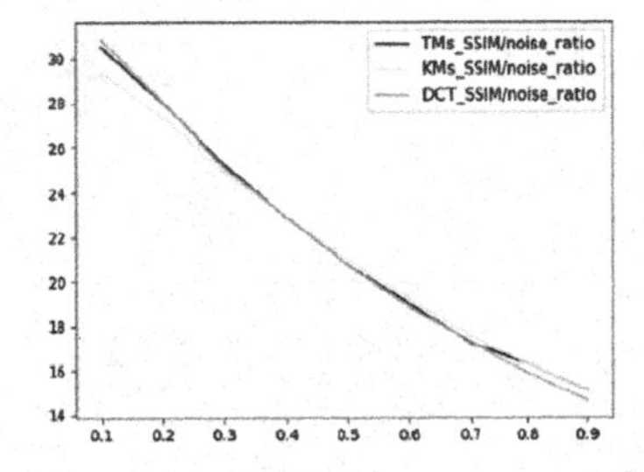

Fig. 6. Selected coefficients on the left and SSIM on the right for TMs, KMs and DCT in Lena.

Table 4. TMs, KMs, DCT denoising performance in Baboon.

Noise ratio	PSNR-TMs	PSNR-KMs	PSNR-DCT	SSIM- TMs	SSIM- KMs	SSIM- DCT
0.1	24.89	24.23	24.84	0.79	0.76	0.79
0.2	24.03	23.47	23.96	0.73	0.71	0.73
0.3	22.74	22.42	22.63	0.65	0.65	0.65
0.4	21.33	21.55	21.21	0.57	0.56	0.57
0.5	19.98	19.89	19.88	0.51	0.49	0.51

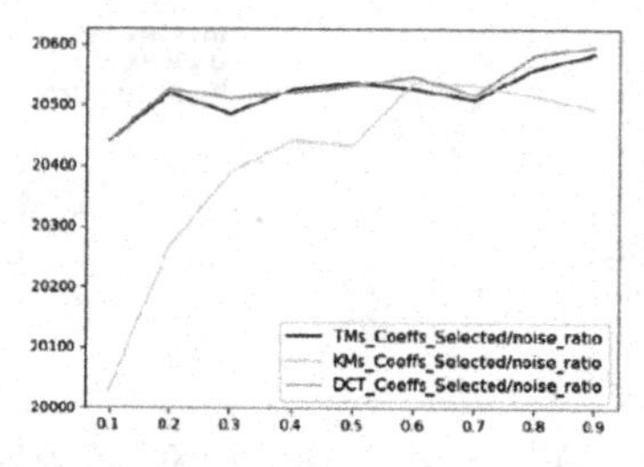

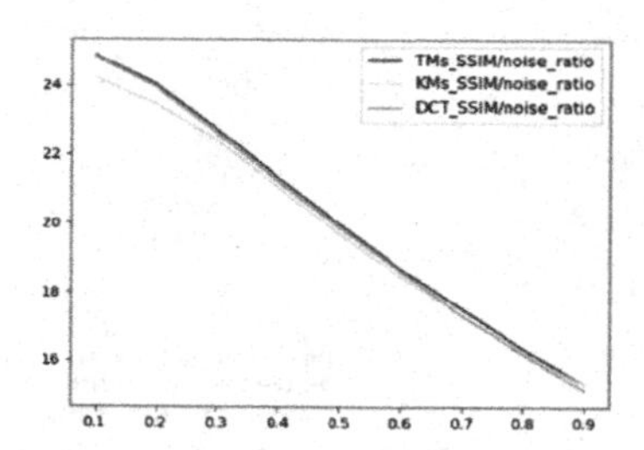

Fig. 7. Selected coefficients on the left and SSIM on the right for TMs, KMs and DCT in Baboon.

Table 5. TMs, KMs, DCT denoising performance in Pirate.

Noise ratio	PSNR-TMs	PSNR-KMs	PSNR-DCT	SSIM-TMS	SSIM-KMs	SSIM-DCT
0.1	27.01	26.16	27.04	0.85	0.82	0.85
0.2	25.66	25.21	25.68	0.77	0.76	0.77
0.3	23.87	23.72	23.18	0.69	0.68	0.68
0.4	21.98	22.03	21.81	0.58	0.59	0.58
0.5	20.07	20.25	19.83	0.48	0.51	0.49

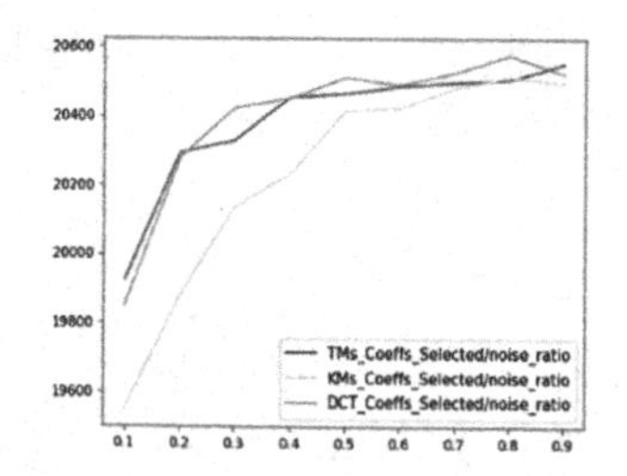

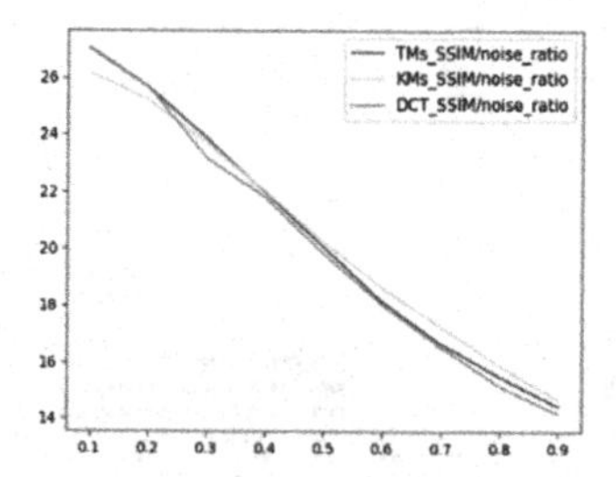

Fig. 8. Selected coefficients on the left and SSIM on the right for TMs, KMs and DCT in Pirate.

Lena DCT achieves the highest PSNR in low noise ratios with small differences with TMs, where it takes the leading place for higher noise ratio than DCT, a conclusion that is confirmed in all cases. Same as previous in the last two cases TMs achieve more sparse results than DCT. Finally, an important notice is that TMs are more robust than DCT as the noise level increases.

Krawtchouk moments and DCT achieve the sparsest solutions in all cases. KMs are not achieved high scores in PSNR and SSIM like DCT in low noise ratio but we can come to the conclusion that same with TMs for higher noise ratios KMs are bringing almost equal metrics with DCT and even better in some cases. Finally, it is important to mention that according to all figures, image moments in comparison with DCT are bringing better metrics for higher noise ratios and the sparsest solutions in all levels of noise for all benchmark images. In the following Figs. 9, 10, 11, 12 and 13, the denoised images for the case of 20% noise ratio for all benchmark images are presented.

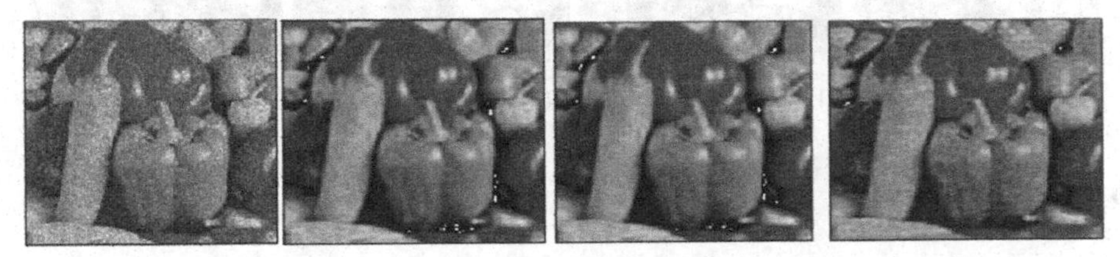

Fig. 9. Peppers: From left to right noisy image, DCT, TMs, and KMS denoised images.

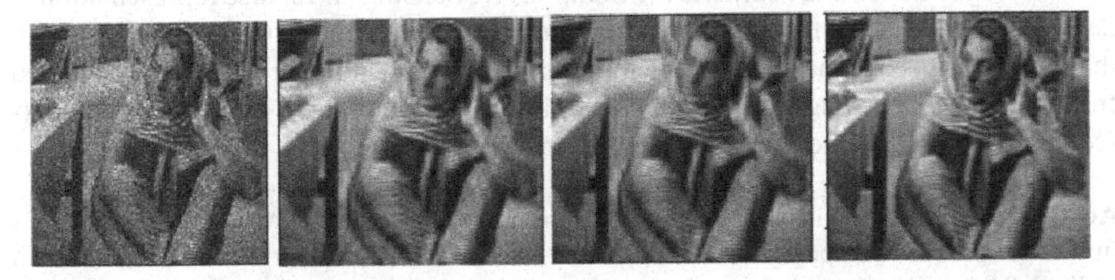

Fig. 10. Barbara: From left to right noisy image, DCT, TMs, and KMS denoised images.

Fig. 11. Lena: From left to right noisy image, DCT, TMs, and KMS denoised images.

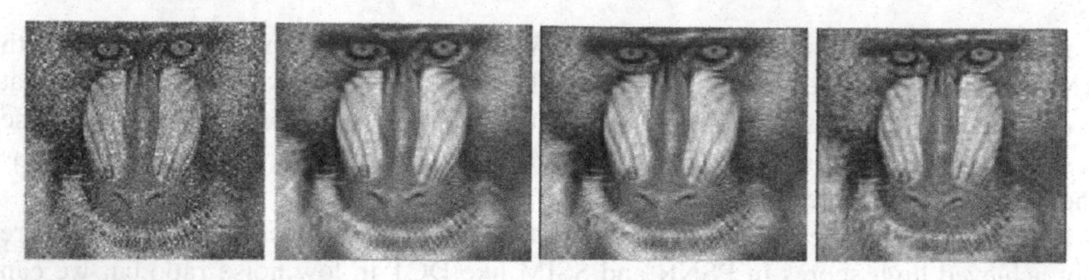

Fig. 12. Baboon: From left to right noisy image, DCT, TMs, and KMS denoised images.

Fig. 13. Pirate: From left to right noisy image, DCT, TMs, and KMS denoised images.

6 Conclusions

The previous study proved that moment transform based on Tchebichef and Krawtchouk families are bringing more sparse solutions than DCT with almost equal metrics. As a result, image moments are suitable for Compressive Sensing and sparse representation in denoising applications from the point of strong and convenient features. In this context, the interest arises on what other problems of this theory can moment transform be applied from dictionary learning to other domains. Moreover, fractional or quaternion moments recently proposed in the literature can be applied.

Acknowledgements. This work was supported by the MPhil program "Advanced Technologies in Informatics and Computers", hosted by the Department of Computer Science, International Hellenic University, Greece.

References

1. Donoho, D.L.: Compressed sensing. IEEE Trans. Inf. Theory **52**(4), 1289–1306 (2006). https://doi.org/10.1109/TIT.2006.871582
2. Candès, E.J.: Compressive sampling. Int. Congress Math. ICM **2006**(3), 1433–1452 (2006). https://doi.org/10.4171/022-3/69
3. Boulanger, S., et al.: A compressive-sensing based watermarking schme for sparse image tampering identification. In: Politecnico di Milano Dip. Elettronica e Informazione - Italy Universit 'a di Siena Dip. Ingegneria dell' Informazione - Italy. System **59**(12), 1265–1268 (2009)
4. Hua, G., Xiang, Y., Bi, G.: When compressive sensing meets data hiding. IEEE Signal Process. Lett. **23**(4), 473–477 (2016). https://doi.org/10.1109/LSP.2016.2536110

5. Candès, E., Romberg, J.: Sparsity and incoherence in compressive sampling. Inverse Prob. **23**(3), 969–985 (2007). https://doi.org/10.1088/0266-5611/23/3/008
6. Guo, Z., Sun, J., Zhang, D., Wu, B.: Adaptive perona-malik model based on the variable exponent for image denoising. IEEE Trans. Image Process. **21**(3), 958–967 (2012). https://doi.org/10.1109/TIP.2011.2169272
7. Elad, M.: On the origin of the bilateral filter and ways to improve it. IEEE Trans. Image Process. **11**(10), 1141–1151 (2002). https://doi.org/10.1109/TIP.2002.801126
8. Starck, J.L., Candès, E.J., Donoho, D.L.: The curvelet transform for image denoising. IEEE Trans. Image Process. **11**(6), 670–684 (2002). https://doi.org/10.1109/TIP.2002.1014998
9. Candes, E.J., Tao, T.: Decoding by linear programming. IEEE Trans. Inf. Theory **51**(12), 4203–4215 (2005). https://doi.org/10.1109/TIT.2005.858979
10. Ansari, R.A., Budhhiraju, K.M.: A comparative evaluation of denoising of remotely sensed images using wavelet, curvelet and contourlet transforms. J. Indian Soc. Remote Sens. **44**(6), 843–853 (2016). https://doi.org/10.1007/s12524-016-0552-y
11. Starck, J.L., Fadili, J., Murtagh, F.: The undecimated wavelet decomposition and its reconstruction. IEEE Trans. Image Process. **16**(2), 297–309 (2007). https://doi.org/10.1109/TIP.2006.887733
12. Wang, F., Wang, S., Hu, X., Deng, C.: Compressive sensing of image reconstruction based on shearlet transform. In: Advances in Intelligent and Soft Computing, 125 AISC, pp. 445–451 (2012). https://doi.org/10.1007/978-3-642-27329-2_61
13. Dragotti, P.L.: Directionlets : anisotropic multidirectional. Image (Rochester, N.Y.) **15**(7), 1916–1933 (2006)
14. Papakostas, G.A.: Improving the recognition performance of moment features by selection. In: Stańczyk, U., Jain, L.C. (eds.) Feature Selection for Data and Pattern Recognition. SCI, vol. 584, pp. 305–327. Springer, Heidelberg (2015). https://doi.org/10.1007/978-3-662-45620-0_13
15. Papakostas, G.A.: Over 50 Years of Image Moments and Moment Invariants. In: Papakostas, G.A. (ed.) Moments and Moment Invariants - Theory and Applications. Gate to Computer Science and Research (GCSR), vol. 1, pp. 3–32 (2014). https://doi.org/10.15579/gcsr.vol1.ch1
16. Kadir, A., Nugroho, L.E., Susanto, A., Insap Santosa, P.: Experiments of zernike moments for leaf identification. J. Theor. Appl. Inf. Technol. **41**(1), 82–93 (2012)
17. Mukundan, R., Ong, S.H., Lee, P.A.: Image analysis by Tchebichef moments. IEEE Trans. Image Process. **10**(9), 1357–1364 (2001). https://doi.org/10.1109/83.941859
18. Yap, P.T., Paramesran, R., Ong, S.H.: Image analysis by Krawtchouk moments. IEEE Trans. Image Process. **12**(11), 1367–1377 (2003). https://doi.org/10.1109/TIP.2003.818019
19. Zhu, H., Shu, H., Zhou, J., Luo, L., Coatrieux, J.L.: Image analysis by discrete orthogonal dual Hahn moments. Pattern Recogn. Lett. **28**(13), 1688–1704 (2007). https://doi.org/10.1016/j.patrec.2007.04.013
20. Papakostas, G.A., Koulouriotis, D.E., Karakasis, E.G., Tourassis, V.D.: Moment-based local binary patterns: a novel descriptor for invariant pattern recognition applications. Neurocomputing **99**(January), 358–371 (2013). https://doi.org/10.1016/j.neucom.2012.06.031
21. Maliamanis, T., Papakostas, G.A.: DOME-T: adversarial computer vision attack on deep learning models based on Tchebichef image moments. In: Proceeding of SPIE 11605, 13th International Conference on Machine Vision, 116050D (4 January 2021). https://doi.org/10.1117/12.2587268
22. Tziridis, K., Kalampokas, T., Papakostas, G.A.: EEG signal analysis for seizure detection using recurrence plots and tchebichef moments. In: IEEE 11th Annual Computing and Communication Workshop and Conference (CCWC), pp. 0184–0190 (2021). https://doi.org/10.1109/CCWC51732.2021.9376134

Classification of Toxic Ornamental Plants for Domestic Animals Using CNN

Sara S. Satake[1], Rodrigo Calvo[1], Alceu S. Britto Jr.[2], and Yandre M. G. Costa[1(✉)]

[1] State University of Maringa (UEM), Maringá, Brazil
{pg402922,rcalvo}@uem.br, yandre@din.uem.br
[2] Pontifical Catholic University of Parana (PUCPR), Curitiba, Brazil
alceu@ppgia.pucpr.br

Abstract. Veterinary medicine emphasizes accidents caused by toxic plants with domestic animals as an extremely important topic, as the right diagnosis can be crucial for the affected animal. In this work, we propose the classification of toxic ornamental plants, according to nine different categories, using five widely-known CNN architectures, namely: DenseNet, ResNet, VGG16, VGG19 and Xception. The rationale behind it is that the automatic identification of these types of plant can be a useful tool to help in the prevention of those accidents. The authors have carefully curated a database to support the development of this work, collecting images available on the Pinterest website, and also performing some important data pre-processing. This database was also made available as a contribution of this work. Transfer learning was employed by taking advantage of feature learned from the ImageNet dataset. We also analyzed the heat maps generated by the Layer-wise Relevant Propagation method, which allowed to observe the individual behavior of the best and worst architectures. The best performance was achieved using DenseNet, with an accuracy of 97.67%. That model managed to generalize very well, even to deal with noisy images, which are frequent in photos of decorative environments.

Keywords: Plant classification · Toxic ornamental plants for animals · Convolutional Neural Networks · Computer vision · Machine learning · Pattern recognition · Layer-wise Relevant Propagation

1 Introduction

Plant cultivation is a common practice in most households, whether indoors or outdoors, as they are important decorative objects, with many varieties. Its popular use is an art passed down orally through information transmitted by generations [2]. However, among the species appreciated, many have toxic potential, which can reflect serious consequences when their active principle is introduced into the organisms of living beings [15].

G. Rozinaj and R. Vargic (Eds.): IWSSIP 2021, CCIS 1527, pp. 108–120, 2022.
https://doi.org/10.1007/978-3-030-96878-6_10

Ornamental plants are those that are able to awake various stimuli through their characteristics, whether by color, texture, size, shape, or harmonic composition according to the context [1]. Such plants are considered toxic when they have substances that can alter the functional-organic set due to organic incompatibility in certain metabolisms, causing different biological reactions [19]. The ingestion of plants is a recurrent habit of domestic animals, especially dogs and cats in different places, public or private, such as gardens, interiors of houses and parks; motivated by various impulses, including curiosity, monotony, age of the animal and change of environment [11].

Veterinary medicine emphasizes the importance of intoxication of small animals by these plants, as the symptoms can often be confused with infectious or parasitic diseases. Therefore, the correct diagnosis must specify correctly whether there was contact with a possible toxic plant [18].

Convolutional Neural Networks (CNNs) have been receiving a great prominence in image recognition, allowing its wide use due to large repositories of public images (e.g. ImageNet), and high-performance computing systems (e.g. GPUs and clusters) [16]. Therefore, its use for classification is increasingly widespread.

Thus, the objective of our work is to implement the classification of nine categories of toxic ornamental plants species for domestic animals, using five CNN architectures: DenseNet, ResNet, VGG16, VGG19 and Xception.

To carry out this work, we used data published by the Toxicological Information Center of the Brazilian state of Rio Grande do Sul (CIT/RS)[1]. According to CIT/RS, in the period from 2001 to 2009, the preeminent accidents involving poisoning of small animals through toxic plants were caused by: Castor Bean (*Ricinus communis*), Peace Lily (*Spathiphyllum wallisii*), Pothos (*Epipremnum pinnatum*), Snake Plant (*Dracaena trifasciata*) and Dumb Cane (*Dieffenbachia seguine*) [5]. One of the major compounds of these plants are the calcium oxalate crystals. The main symptoms of this intoxication include oral irritation with abundant salivation, vomiting, colic, bloody diarrhea, depression, weakness, and in some cases, it can cause the death of the animal by poisoning [23].

We also considered the 2019 article published (see Footnote 1) by CIT/RS, which reports exposures of domestic animals to toxic plants that year. In addition to the plants mentioned above, we included Anthurium (*Anthurium spp.*), Rue (*Ruta graveolens*), Calla Lily (*Calla aethiopica*) and Swiss Cheese Plant (*Monstera delicious*) into the dataset as well. Through the information obtained by CIT/RS, we built a database containing 900 images of the aforementioned plants. Data pre-processing and data augmentation were also performed, and the best performance was obtained using the DenseNet architecture, reaching 97.67% of accuracy.

2 Related Works

Classification involving plants using deep learning is a well-accepted practice for different purposes, as it can be seen in the recurring solutions to the challenges

[1] http://www.cit.rs.gov.br/.

launched in The Plant Identification Task of LifeCLEF, in which the participants are asked to identify images of plants taken from different environments [17].

Ghazi et al. [6] used the GoogleNet, AlexNet and VGGNet architectures to identify plant species in photographs, and then, they measured the performance of these architectures. In addition, the image database was increased using data augmentation. The authors performed a combination of GoogleNet and VGGNet architectures, obtaining a better accuracy. The method used by the authors performed better than those originally obtained in LifeCLEF 2015.

Goeau et al. [7] gathered the main works submitted for LifeCLEF 2017, and observed that those using CNNs obtained a better result, especially when submitted to images with noise, demonstrating a good adaptability and generalization in those cases, even with a constant number of training iterations.

The work of Lee et al. [9] addressed the identification of 44 plant species using a pre-trained CNN for automatic learning of leaf characteristics, avoiding the use of hand-crafted methods. This model proved to be successful to address that task and reached an accuracy of 99.5%, being superior to conventional learning methods, based on the use of handcrafted features and SVM classifier.

Yalcin and Razavi [21] proposed a CNN architecture to identify 16 types of plants common in agriculture. In the same work, SVM-based classifiers were implemented, using resources such as LBP and GIST. The CNN model obtained the highest accuracy rate, around 97.47%; in contrast, the best accuracy obtained by a SVM classifier was 89.94%. Thus, CNN proved to be a good method to achieve the expected objective.

Finally, Nakahata et al. [12] experimented five different CNN architectures aiming at performing bonsai style classification, taking into account seven different categories. The best result (i.e. 0.896 of F-Measure) was obtained using the VGG19 with features learned on the ImageNet.

3 Proposed Method

The steps followed to reach the objectives of this work are described in the following subsections.

3.1 Database Creation

As the focus of our work are toxic ornamental plants, we needed to create a database containing the nine classes identified in the problem. The acquisition of the images was carried out through the Pinterest[2] platform, which brings together a large number of users who organize images around a certain theme, resulting in a rich collection of metadata [24]. Each class received 100 images, totaling 900 images. Aiming at circumventing copyright restrictions, the content of the database can be accessed through the website designed especially for this work[3], where the URLs of the images used are listed, and not the images themselves. Figure 1 presents one example of image from each class.

[2] https://pinterest.com/.

[3] https://sites.google.com/view/toxic-ornamental-plants.

3.2 Data Pre-processing

The images selected to compose this database were obtained manually. In addition, the images were pre-processed according to the steps described following.

Fig. 1. Examples of image for each class.

In the first step, we resized all the images in order to standardize them. So, all samples received the dimensions of 224×224 pixels, since it is a value commonly used in CNN architectures.

As the database was created manually without pre-filtering repeated images, we implemented mechanisms to ensure that all samples were different from each other. Aiming to guarantee a good level of integrity for the base, we used the techniques Mean Squared Error (MSE) and Structural Similarity (SSIM) to detect similar images. In some classes, the similarity degree was very high due to the positioning and background, being necessary to submit them into two verifiers to guarantee the proof of the images distinction.

MSE is a popular similarity index for measuring distance in images. For two images x and y, sized $N \times M$, the MSE can be calculated by Eq. 1 [14].

$$MSE = (\frac{1}{NM})\sum_{i=1}^{M}\sum_{j=1}^{N}(x(i,j) - y(i,j))^2 \tag{1}$$

The lower the MSE result, the more similar the images are.

SSIM provides good image quality prediction performance, being widely used in situations involving image distortion. In a spatial domain between two image patches $x = x_i|i = 1, ...M$ and $y = y_i|i = 1...M$, the SSIM index is defined in the Eq. 2 [14].

$$S(x,y) = \frac{(2\mu_x\mu_y + C_1)(2\delta_{xy} + C_2)}{(\mu_x^2 + \mu_y^2 + C_1)(\delta_x^2 + \delta_y^2 + C_2)} \tag{2}$$

where C_1 and C_2 are two small positive constants, $\mu_x = \frac{1}{M}\sum_{i=1}^{M} x_i$, $\delta_x^2 = \frac{1}{M}\sum_{i_1}^{M}(x_i - \mu_x)^2$, $\delta_{xy} = \frac{1}{M}\sum_{i=1}^{M}(x_i - \mu_x)(y_i - \mu_y)$. The SSIM index value 1 is reached if and only if x and y are identical.

Fig. 2. Example of similar images and the MSE and SSIM values between them.

We discarded and replaced all images with MSE index below 3000, and SSMIM index above 8.5. Figure 2 presents an example of two images of the same class, and the similarity degree between them using the MSE and SSIM indexes.

3.3 Data Augmentation

Due to the small amount of images in the database, we used the data augmentation technique to overcome this problem and avoid over-fitting.

The problem of over-fitting reduces the ability of CNNs to generalize unsupervised data. Thus, the data augmentation process was necessary to enrich the training set, adding new samples in their respective classes, with transformations such as scale, zoom and translations [20]. Ten new images were generated for each original image from each class, totaling a total set of 9900 images. Figure 3 shows transformations of zoom range, width shift range and height shift of 0.3, horizontal flip, vertical flip and 90° rotation range performed during the data augmentation process, which may occur simultaneously.

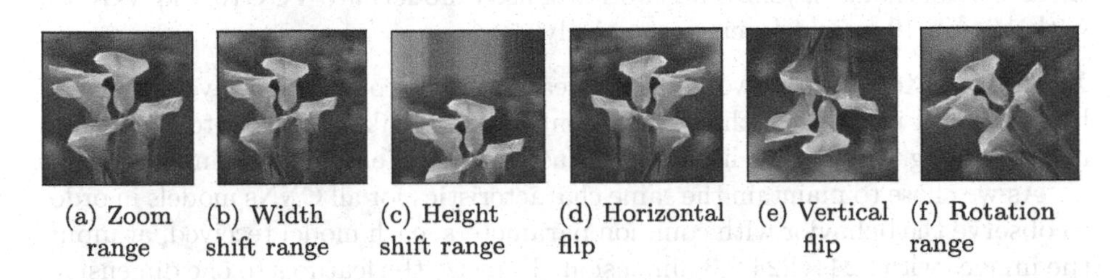

(a) Zoom range (b) Width shift range (c) Height shift range (d) Horizontal flip (e) Vertical flip (f) Rotation range

Fig. 3. Examples of data augmentation.

3.4 Transfer Learning

All CNN architectures were submitted to the transfer learning method using the ImageNet[4] database. It consists of about 15 million labeled images, separated into 22000 categories [10]. Since the classes presented in this work are composed of popular plants, this procedure was a strong ally for pre-training the CNNs models, and allowed them to better generalize their classifications in a smaller number of epochs.

3.5 Convolutional Neural Networks

CNN architectures are extensions of deep learning of artificial neurons [21]. They consist of Multi-hidden Layer Perceptrons (MLPs), with convolutions, pooling, ReLU and fully connected layers. In general, resource maps of the previous layers are convolved with weights that are learnable in a convolutional layer, and, through activation functions, they are fed back to form output resource maps.

In the present work, we used five CNN architectures, namely: DenseNet, ResNet, VGG16, VGG19 and Xception.

DenseNet: In the DenseNet architecture [25], the network structure is progressively hierarchical, resulting in a more generalizable network. Each layer in the network is directly connected to the front layers. To ensure that resource maps are concatenated, they must be consistent. So, the exits from the convolutional layer are the same size as the inlet.

[4] http://www.image-net.org/.

ResNet: The creation of ResNet [3,8] was motivated by counter-intuitive experimental discoveries that indicated that adding more layers increases the chances of errors in training. ResNet allows deeper networks to be trained, maintaining good performance. The models are implemented with double or triple layer skips without linearity (ReLU) and normalization of batches.

VGGNet: The VGGNet architecture [16] is widely used for image classification, obtaining good results for this kind of task. It is a very deep network and uses very small convolution filters (3×3) with a pool layer added after every two or three convolutional layers. The two most used models are VGG16 and VGG19, with 16 and 19 weight layers respectively.

Xception: Xception [4] works as a linear stack of convolution layers that can be separated in depth with residual connections, making it easy to define and modify, using a high-level library, such as Keras or TensorFlow-Slim.

As we chose to maintain the same characteristics for all CNNs models in order to observe the behavior with common parameters, each model received, at input, the images with $224 \times 224 \times 3$ dimension. To resize the features to one dimension, we used a Flatten; after that, we added a dense layer with 256 neurons and a dropout of 0.1. The activation function chosen was ReLU, because it is fast and it does not harm the performance. The output layer received nine neurons (one for each class), along with the Softmax activation function. To calculate the loss, we used the categorical crossentropy function, considering the presence of many classes. We implemented the Stochastic Gradient Descent (SGD) optimizer as it is one of the most popular and faster among the best known techniques, and also the Nesterov-accelerated Adaptive Moment Estimation (Nadam); both with a 10^{-4} learning rate.

3.6 Layer-wise Relevance Propagation (LRP)

To analyze the results of the best architectures, we used the Layer-wise Relevance Propagation (LRP) decomposition method. This technique helps to visually explain network predictions individually. Its function is to decompose the predicted probability of a specific area into a set of relevant pixels, more details in [22].

4 Experimental Results and Discussion

In order to evaluate the performance of each neural network, the five models were implemented individually, and we calculated their results considering a ten-folds cross-validation. The metrics used were recall, precision, F1-score, in addition to accuracy, and the Macro-averaging precision. The division of the image set used for training and validation is available on the website (see Footnote 3) created by the authors. All models were trained in ten epochs.

Initially, we implemented the SGD optimizer due to its greater popularity and its speed in updating the parameters for each training sample [13]. However,

Table 1. Results using SGD optimizer

	Precision (%)	Recall (%)	F1-score (%)	Accuracy(%)
DenseNet	**85.56**	**85.56**	**85.55**	**85.60**
ResNet	23.30	25.00	22.42	23.30
VGG16	49.78	47.88	48.15	49.78
VGG19	49.56	47.20	47.93	49.56
Xception	74.00	74.40	73.82	74.00

we noticed that the results obtained using the SGD did not exceed 85.6%, as it can be seen in Table 1.

Searching for better results, we decided to evaluate the Nadam optimizer. In this way, we have successfully achieved better classification rates. The new results are shown in Table 2.

Table 2. Results using Nadam optimizer

	Precision (%)	Recall (%)	F1-score (%)	Accuracy (%)
DenseNet	**97.67**	**97.67**	**97.70**	**97.67**
ResNet	53.44	61.73	52.86	53.44
VGG16	92.55	92.61	92.54	92.56
VGG19	94.44	94.46	94.42	94.44
Xception	94.22	94.22	94.21	94.22

Considering the results obtained using the Nadam optimizer, we also generated the boxplot for each architecture, shown in Fig. 4.

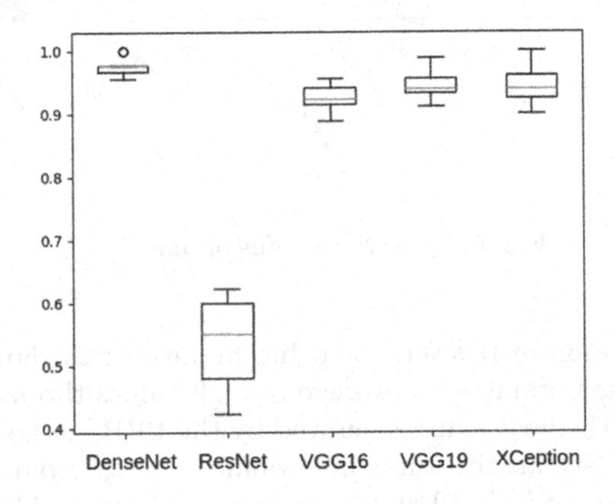

Fig. 4. Boxplots created from the five CNN architectures

As noted, the best architecture for the proposed objective was DenseNet, achieving 97.67% of total accuracy. Its representation in the boxplot exhibits the high values obtained through folds; there was an outlier occurrence, where the accuracy of one of the folds reached 100%. Figure 5 shows the confusion matrix for this model.

We noticed that the class that had more errors was Pothos, with a subtotal of 93 hits; its erroneous classification was given by the Anthurium class in two occurrences. The best performance happened in the Snake Plant class, where there was the maximum number of hits. In fact, the samples for this class had the lowest incidence of noise, and a better standardization of images. In addition, the shape of this plant is very peculiar and quite distinct from the other plants analyzed in the work.

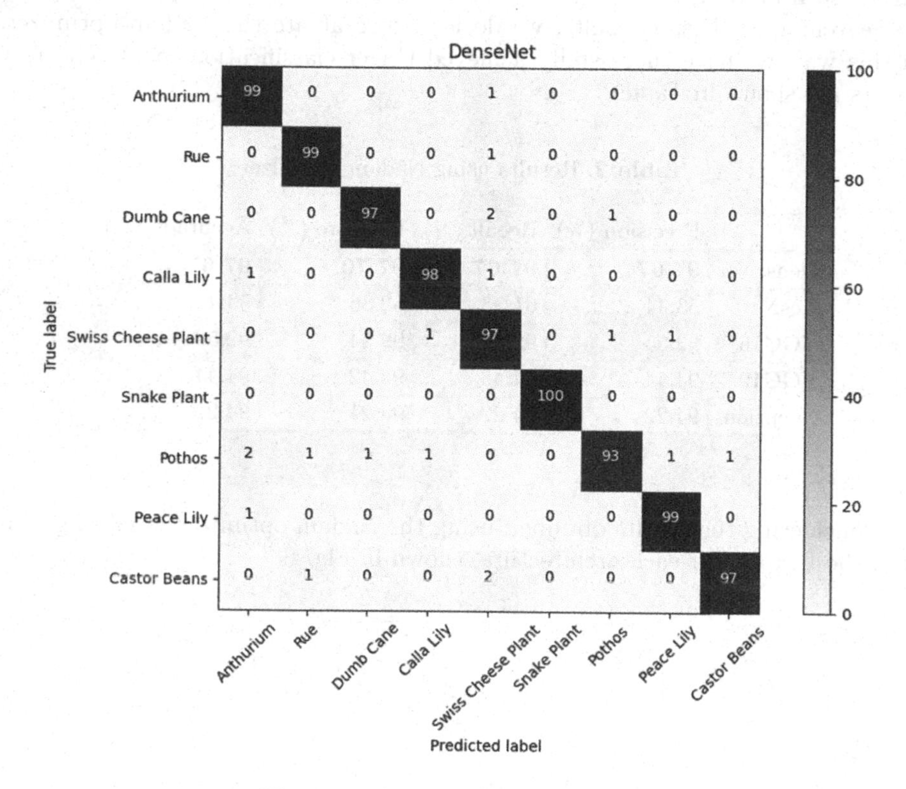

Fig. 5. DenseNet confusion matrix.

With the exception of ResNet, the other architectures showed low variance in their accuracy, maintaining an average of high values through their folds.

Figure 6 shows the heat maps generated by the LRP method for the ResNet, VGG19 and DenseNet architectures in a random image from each class of the data set. In the case of VGG19 and DenseNet, both were able to map positive pixels relevant to the shape of the plant. It is possible to observe, through the

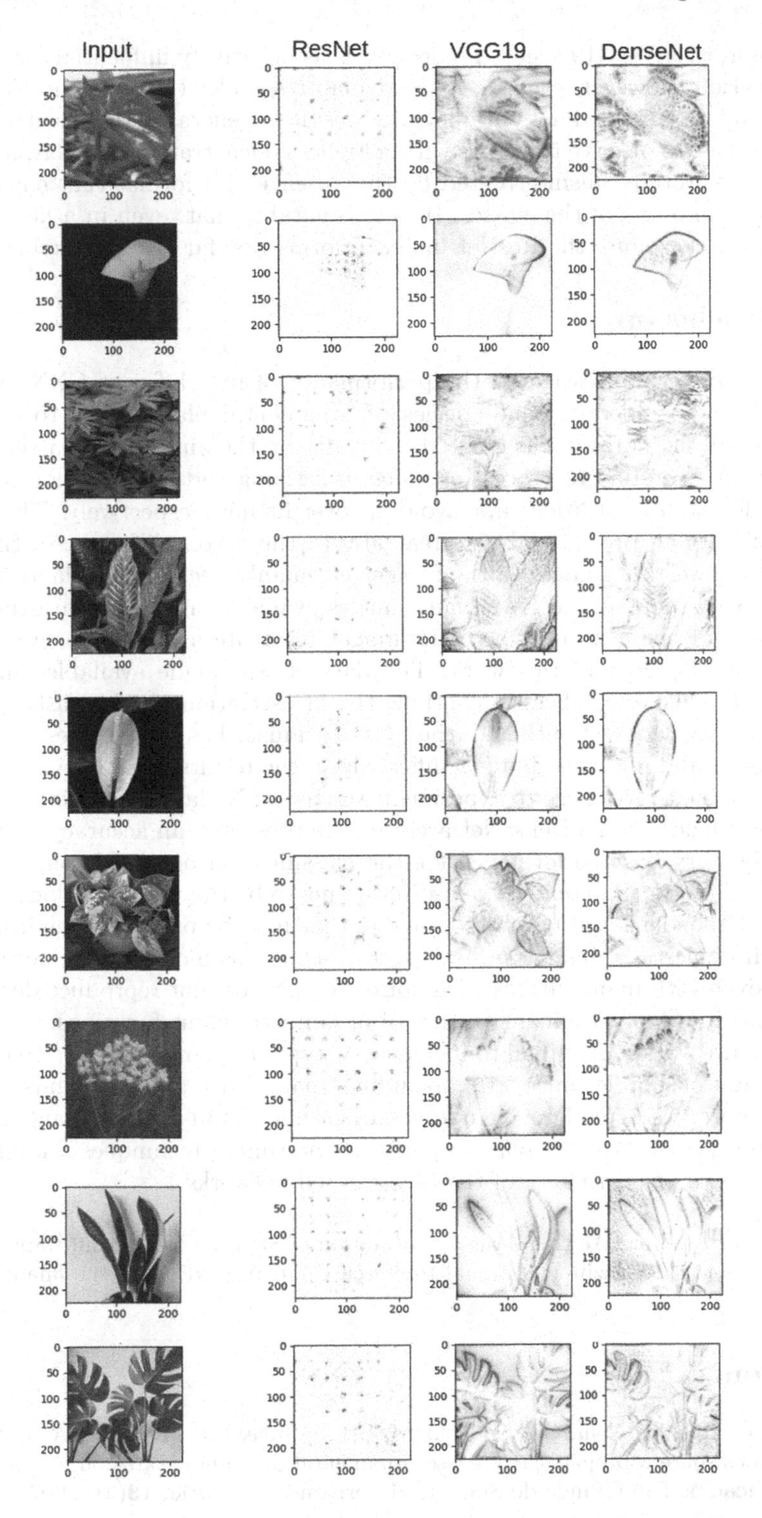

Fig. 6. Heat maps created with LRP method for ResNet, VGG19 and DenseNet.

blue color, that VGG19 accused more pixels as negatively influential for a future classification. However, even so, its heat map resembles that of DenseNet.

ResNet, in turn, did not obtain a good result to generalize the relevant pixels. We were able to observe the areas with a higher concentration of pixels, however, they can be totally misinterpreted by the absence of a format that could characterize it into one of the classes. It is also notable that, even in a neutral and noiseless background, the ResNet underperformed by far the best architectures.

5 Conclusion

In this paper, we investigated the performance of five different CNN architectures aiming to identify nine species of ornamental plants toxic to domestic animals. As the dataset was created manually by the authors, it was necessary to perform a careful pre-processing, such as resizing and checking non-repeated images for standardization and avoiding over-fitting, respectively. The use of transfer learning proved to be a great ally to the accomplishment of this task when using weights trained on the ImageNet, mainly because the plants in question are very popular and have many images available on online platforms, such as Pinterest, where we obtained the images to create a fully exclusive dataset for the development of this work. The dataset was made available aiming to encourage other researchers to continue the investigations in this task.

As we have noticed, although the SGD optimizer has a large presence in the literature, it did not contribute significantly to the results obtained in this work, in general. Thus, we chose to experiment the use of Nadam, increasing the accuracy of all models. The DenseNet architecture presented an accuracy of 97.67%, being the very best performance for the classification on the dataset we have created. Thus, in this context, DenseNet seems to be the most suitable model to carry out the identification of toxic plants that may be present in environments in which domestic animals can have access, since this model has achieved good results even with many images with noise, i.e., photos that reproduce decorative spaces, as it is a predominant feature of ornamental plant functions.

In future works, we intend to include new types of plants, and also to increase the amount of samples in general, including images with different kinds of noise. Furthermore, we intend to include some specific details of the plant, such as individual leaves. We also plan to perform fine-tuning techniques, individually, to improve the performance of the investigated networks.

Acknowledgment. We thank the Coordination of Superior Level Staff Improvement (CAPES) and the Brazilian National Research Council (CNPq) for the financial support.

References

1. Barroso, C.M., Klein, G.N., de Barros, I.B., Franke, L.B., Delwing, A.B.: Considerações sobre a propagação e o uso ornamental de plantas raras ou ameaçadas de extinção no Rio Grande do Sul, Brasil. Ornamental Hortic. **13**(2) (2007)

2. de Mello Botelho, J., do Nascimento Lamano-Ferreira, A.P., Ferreira, M.L.: Prática de cultivo e uso de plantas domésticas em diferentes cidades brasileiras. Ciência Rural **44**(10), 1810–1815 (2014)
3. Chen, Z., Xie, Z., Zhang, W., Xu, X.: ResNet and model fusion for automatic spoofing detection. In: INTERSPEECH, pp. 102–106 (2017)
4. Chollet, F.: Xception: deep learning with depthwise separable convolutions. In: Proceedings of the IEEE CVPR, pp. 1251–1258 (2017)
5. Conceição, J.L.S., Ortiz, M.A.L.: Intoxicação domiciliar de cães e gatos. Revista UNINGÁ Rev. **24**(2), 59–62 (2015)
6. Ghazi, M.M., Yanikoglu, B., Aptoula, E.: Plant identification using deep neural networks via optimization of transfer learning parameters. Neurocomputing **235**, 228–235 (2017)
7. Goeau, H., Bonnet, P., Joly, A.: Plant identification based on noisy web data: the amazing performance of deep learning. In: LifeCLEF 2017 (2017)
8. He, K., Zhang, X., Ren, S., Sun, J.: Deep residual learning for image recognition. In: Proceedings of the IEEE Conference on Computer Vision and Pattern Recognition, pp. 770–778 (2016)
9. Lee, S.H., Chan, C.S., Wilkin, P., Remagnino, P.: Deep-plant: plant identification with convolutional neural networks. In: 2015 IEEE International Conference on Image Processing (ICIP), pp. 452–456. IEEE (2015)
10. Marmanis, D., Datcu, M., Esch, T., Stilla, U.: Deep learning earth observation classification using ImageNet pretrained networks. IEEE Geosci. Remote Sens. Lett. **13**(1), 105–109 (2015)
11. Martins, D.B., Martinuzzi, P.A., Sampaio, A.B., Viana, A.N.: Plantas tóxicas: uma visão dos proprietários de pequenos animais. Arquivos de Ciências Veterinárias e Zoologia da UNIPAR **16**(1), 11–17 (2013)
12. Nakahata, G.H.S., Constantino, A.A., Costa, Y.M.G.: Bonsai style classification: a new database and baseline results. In: 2020 IEEE International Symposium on Multimedia (ISM), pp. 104–110. IEEE (2020)
13. Ribeiro, A.M., et al.: Um estudo comparativo entre cinco métodos de otimização aplicados em uma rnc voltada ao diagnóstico do glaucoma. Revista de Sistemas e Computação-RSC **10**(1), 122–130 (2020)
14. Sampat, M.P., Wang, Z., Gupta, S., Bovik, A.C., Markey, M.K.: Complex wavelet structural similarity: a new image similarity index. IEEE Trans. Image Process. **18**(11), 2385–2401 (2009)
15. Santos, C.: Plantas ornamentais tóxicas para cães e gatos presentes no Nordeste do Brasil. Medicina Veterinária (UFRPE) **7**(1), 11–16 (2013)
16. Simonyan, K., Zisserman, A.: Very deep convolutional networks for large-scale image recognition. arXiv preprint arXiv:1409.1556 (2014)
17. Sünderhauf, N., McCool, C., Upcroft, B., Perez, T.: Fine-grained plant classification using convolutional neural networks for feature extraction. In: CLEF (Working Notes), pp. 756–762 (2014)
18. Tenedini, V., Dos Anjos, B.L., Mafra, J.R.: Plantas ornamentais tóxicas para cães e gatos. Anais do Salão Internacional de Ensino, Pesquisa e Extensão **7**(3) (2016)
19. Vasconcelos, J., de Pontes Vieira, J.G., de Pontes Vieira, E.P.: Plantas tóxicas: conhecer para prevenir. Revista Científica da UFPA **7**(1), 1–10 (2009)
20. Xu, Q., Zhang, M., Gu, Z., Pan, G.: Overfitting remedy by sparsifying regularization on fully-connected layers of CNNs. Neurocomputing **328**, 69–74 (2019)
21. Yalcin, H., Razavi, S.: Plant classification using convolutional neural networks. In: 2016 5th International Conference on Agro-Geoinformatics (Agro-Geoinformatics), pp. 1–5. IEEE (2016)

22. Yang, Y., Tresp, V., Wunderle, M., Fasching, P.A.: Explaining therapy predictions with layer-wise relevance propagation in neural networks. In: 2018 IEEE International Conference on Healthcare Informatics (ICHI), pp. 152–162. IEEE (2018)
23. Zeinsteger, P., Gurni, A.: Plantas tóxicas que afectan el aparato digestivo de caninos y felinos. Revista Veterinaria **15**(1), 35–44 (2004)
24. Zhai, A., et al.: Visual discovery at Pinterest. In: Proceedings of the 26th International Conference on World Wide Web Companion, pp. 515–524 (2017)
25. Zhang, J., Lu, C., Li, X., Kim, H.J., Wang, J.: A full CNN based on DenseNet for remote sensing scene classification. Math. Biosci. Eng **16**(5), 3345–3367 (2019)

Deep Learning-Based Detection of Seedling Development from Indoor to Outdoor

Hadhami Garbouge[1], Pejman Rasti[1,2], and David Rousseau[1](✉)

[1] Université d'Angers, LARIS, UMR INRAe IRHS, 62 Avenue Notre Dame du Lac, 49000 Angers, France
david.rousseau@univ-angers.fr

[2] Department of Computer Science, ESAIP, Angers, France

Abstract. Monitoring plant growth with computer vision is an important topic in plant science. This monitoring can be challenging when plants are located in outdoor conditions due to light variations and other noises. On other hand, there is a lack of annotated datasets available for such outdoor environments to train machine learning algorithms while indoor similar datasets may be more easily available. In this communication, we investigate, for the first time to the best of our knowledge in plant imaging, how to take benefit from model trained in fully controlled environment to build model for an outdoor environment. This is illustrated with a use case recently published for indoor conditions that we revisit and extend. We compare various spatial and spatio-temporal neural network architectures including long-short term memory convolutional neural network, time distributed convolutional neural network and transformer. While the spatio-temporal architectures outperform the spatial one in indoor conditions, the temporal information appears to be degraded by the presence of shadows due to the variation of light in outdoor conditions. We introduce a specific data augmentation and transfer learning approach which enables to reach a performance of 91% of good classifications with very limited effort of annotation.

Keywords: Plant phenotyping · Deep learning · Transfer learning · Data augmentation · CNN-LSTM · Time distributed deep learning · Transformers

1 Introduction

With the breakthrough of deep learning almost ten years ago, the state-of-the-art in most data-driven image processing domains shifted from classical machine learning with hand-crafted features to end-to-end learning [9]. As a consequence, the bottleneck in developing image processing solutions are now related to the annotation of images which is a very time-consuming and error-prone task. This bottleneck is specially important in applied domains of computer vision for which

G. Rozinaj and R. Vargic (Eds.): IWSSIP 2021, CCIS 1527, pp. 121–131, 2022.
https://doi.org/10.1007/978-3-030-96878-6_11

much fewer annotated datasets are publicly available. This is the case for instance with plant imaging [11], the applied domain of this communication. Two main subcommunities are concerned by plant imaging. On one side, plant phenotyping facilities are studying the development of plant in controlled conditions in order to study the genotype-environment interactions and their impact on the observable phenotype [10]. On another side, outdoor conditions are of the highest interest for agricultural practices which can also benefit from computer vision [12]. An open question is if machine learning models trained indoor could be useful to help the outdoor conditions.

Several workaround approaches has been proposed to address the bottleneck of annotation in applied computer vision including the development of ergonomic tools to speed up annotation, data augmentation, transfer learning, generation of simulated images or the use of generative neural networks. These approaches have been applied to the domain of plant imaging and the communication here is in this trend [4,5,14,16]. We recently developed a spatio-temporal deep learning algorithm to monitor the growth of seedlings in a controlled environment from top-view in RGB images [15]. Here, we propose an extension of this work by investigating the possibility to transfer this knowledge to the outdoor environment where the lighting conditions are not controlled and shadow may occur due to the position of the sun or the presence of clouds passing by. This is to the best of our knowledge the first trial of this type in plant imaging.

As most related works to our proposal, one can point that the computer vision community has in recent years addressed the automatic detection and removal of shadows in RGB images with deep learning [2,8,13]. As often encountered when considering the translation of such literature to other application domains some basic practical issues may appear. In the current work, the spatial content and resolution from [2,8,13] are clearly different from the one considered in seedling growth. As a consequence direct transfer learning would very likely fail and would require additional annotated images. Our proposal here is rather to investigate the possible transfer of knowledge from plant observed in indoor conditions to outdoor conditions.

2 Datasets

2.1 Real Indoor and Out Door Data

Two distinct datasets have been produced. The first dataset consists of 449286 images (600 different pots) from red clover (Trifolium pratense) and alfalfa (Medicago sativa) which were captured in a fully controlled environment [15]. This dataset or a pre-processed version of it will serve as the training dataset in this study. The second dataset includes 22212 images (36 different pots) captured from sunflower seedlings in a non controlled-environment (greenhouse). This second dataset serves a testing dataset in this study. Both datasets have been recorded with the frame rate of one image every 15 min. Figure 1 shows an example of each dataset. Both datasets record the first developmental stages of the growth of seedlings. This includes four stages with the soil, the first appearance of the cotyledon (FA), the opening of the cotyledons (OC), the appearance of the first leave (FL).

The objective of the work is to transfer knowledge from a model trained on the first dataset to the second dataset as illustrated in Fig. 2. While the species of both datasets are different they are both dicotyledons so that they share similar shapes at the early stages of development. Moreover, the two cameras share the same spatial resolution. As visible in Fig. 1, the color of the crop observed indoor and outdoor are not exactly the same. This color discrepancy happened to be none critical to transfer knowledge from indoor to outdoor. As done in [15], the plant is filtered from the soil with a standard thresholding approach to avoid any impact on the difference of soil and surrounding background. The challenge in the proposed experiment therefore lay in the presence of shadows which occurs in outdoor environment only.

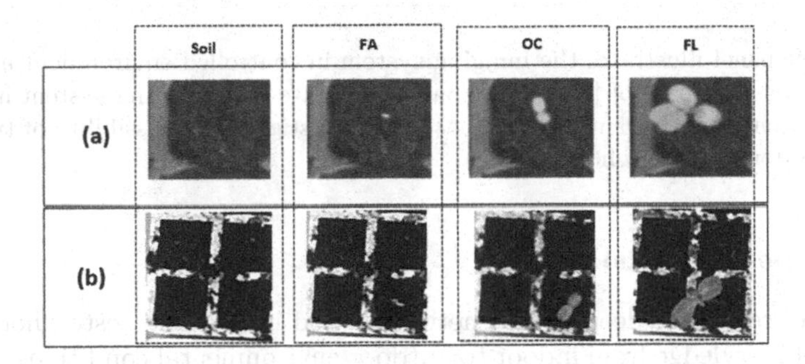

Fig. 1. (a) Images from controlled environment on which seedling development is trained. (b) Images from outdoor environment on which we want to test the trained model. The four developmental stages to be detected are the soil, the first appearance of the cotyledon (FA), the opening of the cotyledons (OC), the appearance of the first leave (FL).

2.2 Simulated Outdoor Data

To simulate images acquired in the outdoor environment from indoor images, we propose an automatic shadow generator as detailed in Algorithm 1. The shadows are randomly positioned by using a thresholded speckle generator [5,7]. All sizes of shadow can be present outdoor. However, only shadows larger than the typical size of seedling organs and smaller than a single plant are expected to impact the detection of seedling development. We adjusted the number of phasors in the speckle generator in order to fit with this prior knowledge and produce shadows corresponding to the maximum area of the seedling (40% of the size of the pot in our training dataset). Modulation of maximum intensity during the day was recorded in the validation dataset. This information was used to adjust the value of the threshold in the algorithm (found to $threshold = 0.5$ in our validation dataset). Each image in the indoor database is then spatially modulated by the generated shadow with a simple multiplication.

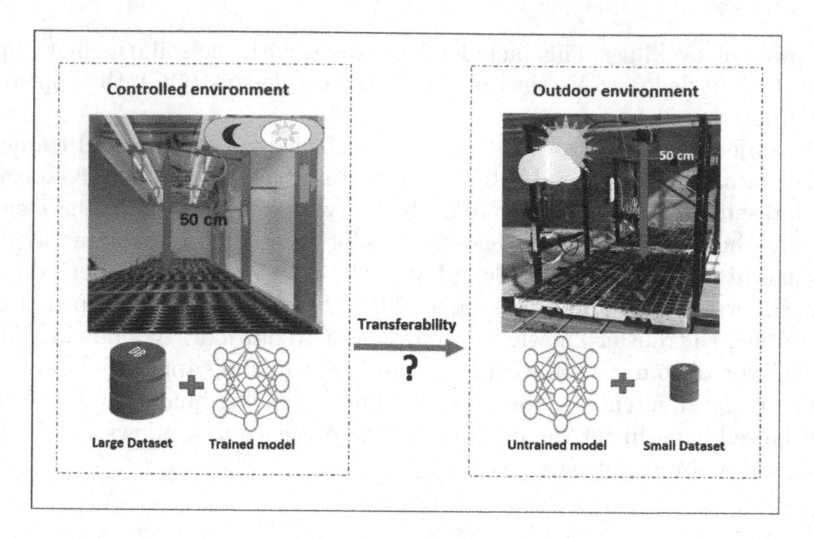

Fig. 2. Left panel illustrates the imaging system in controlled environment associated with the large database of [15]. Right panel illustrates the imaging system in an outdoor environment with a smaller database. We investigate the possibility of transfer of knowledge from left to right panels.

2.3 Proposed Methods

We shortly recall the deep neural networks used in [15] and tested here on the transfer of knowledge from indoor to outdoor environmental conditions. We then extend to other methods, not included in [15] and tested for the first time in plant imaging.

First, we included in [15] a basic CNN architecture performing a 4 classes classification to discriminate between the images of Fig. 1(a). The architecture of CNN is composed of five convolutional layers with filters of size 3×3 and respective numbers of filters 64, 128, 128, 512 and 512 each followed by rectified linear unit (RelU) activations and 2×2 max-pooling; a fully connected layer with 512 units and ReLU activation, a fully connected output layer with 4 classes corresponding to each event and a softmax activation. We use cross-entropy as loss function and adam as optimizer. The architecture optimized for this 4 classes task is visible in Fig. 4 and served as the baseline in [15] since it does not embed any memory about the growth process. We demonstrated in [15] the added value to embed in controlled environment such a memory and demonstrated the superiority of a CNN-LSTM (see Fig. 4.b) by comparison with a sole LSTM architecture (see [15]). The optimal duration of the memory was found to 4 images in [15] corresponding to 1 h of recording.

To further enrich the investigation on memory, we added other neural network architectures. We tested gated recurrent unit (GRU) networks [1], an alternative to LSTM, which has been demonstrated empirically to converge faster. GRU uses two gates: the update gate and the reset gate while there are three

Algorithm 1: Pseudo-code to simulate random shadows

Data: I: Original image, n:number of phases, s:threshold $(0, 1)$.
Result: I_{aug}: Image with shadow
1 l ← height of original image
2 c ← width of original image
3 shadow ← zeros (l,c)
4 Phases ← exp $(2 * \pi * Rand(n, n) * i)$
5 shadow (1:n,1:n) ← Phases
6 shadow ← $|FFTshift(IFFT(shadow))|$
7 shadow ← shadow / (**Max**(shadow)
8 **for** $i \leftarrow 1$ *to* l **do**
9   **for** $j \leftarrow 1$ *to* c **do**
10   **if** $shadow(i, j) < threshold$ **then**
11   shadow(i,j) ← threshold
12 $I_{shadow} = I * shadow$

gates in LSTM. This difference makes GRU faster to train and with better performance than LSTMs on less training data [18]. A last class of neural network dedicated to time series are the transformers. Since their introduction in [17] they have been shown to outperform recurrent neural networks such as LSTM and GRU specially in the field of natural language processing as they do not require that the sequential data be processed in order. Transformers have been shown suitable to process temporal information carried by single pixels in satellite images time series [6,19,20]. Transformers have recently been extended to the process of images [3] where images were analysed as a mosaic of subparts of the original images creating artificial time series. In our case, we directly have meaningful subparts of the original images which corresponds to the field of view of the pots. We, therefore, provide the transformer of [3] with time series of consecutive images of the same pot (we used the same time slot as in the other spatio-temporal methods). We used 32 transformer layers with batch size 64, feed forward layer as classification head layer and the size of our patch size was equal to 89×89 pixels.

The performances of the models proposed in [15] for controlled conditions are recalled in Table 1 in addition to the three new methods added in this communication CNN-GRU, TD-CNN-GRU, Transformer. The performance of the TD-CNN-GRU model and Transformer are found to outperform the other methods in controlled environment. A possible interpretation is that, in the TD-CNN-GRU and Transformer models, time and space are stacked and processed at the same time while CNN-LSTM first processes space and then time in a sequential way. In the following, we investigate how the performances of the methods shown in Table 1 evolve when the models are applied in outdoor environment. For this experiment, we selected the memoryless CNN model and the best time-dependent neural network models: TD-CNN-GRU and Transformer.

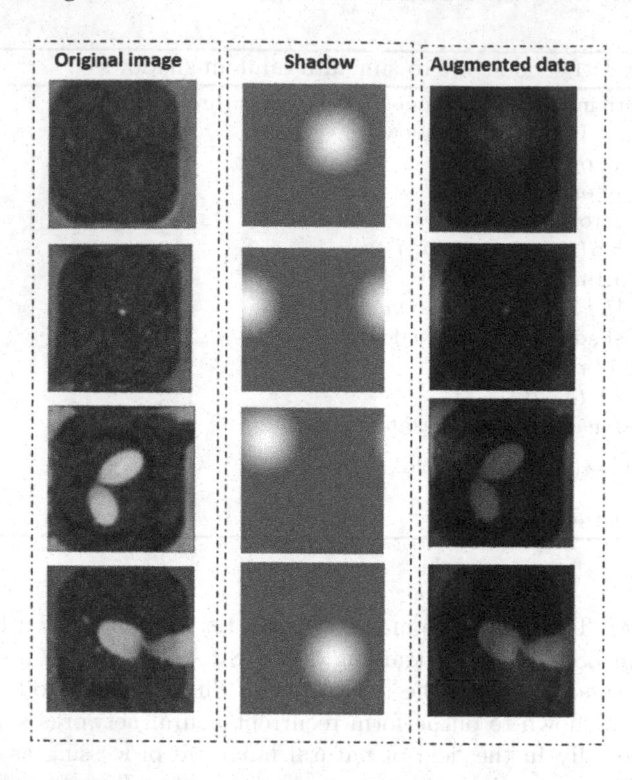

Fig. 3. Example of original indoor images (left), shadows generated with Algorithm 1 (middle) and, indoor images with simulated shadows (right).

Table 1. Tested models in the fully controlled environment. Mean and standard deviation of the accuracy from 5 different trials for each model.

Models	Accuracy
CNN	0.80 ± 0.08
CNN-LSTM	0.90 ± 0.08
CNN-GRU	0.91 ± 0.06
TD-CNN-GRU	0.96 ± 0.01
Transformer	0.92 ± 0.01

3 Results

Several transfers of knowledge has been tested from indoor conditions to outdoor conditions. First, as baseline we have applied a brute transfer where the models trained indoor have directly been applied to predict the outdoor images. The performance with the CNN model, visible in Table 2, shows a clear drop although it does not vanishes to pure randomness. Then, we have used data augmentation based on the simulation of shadows applied on indoor images as

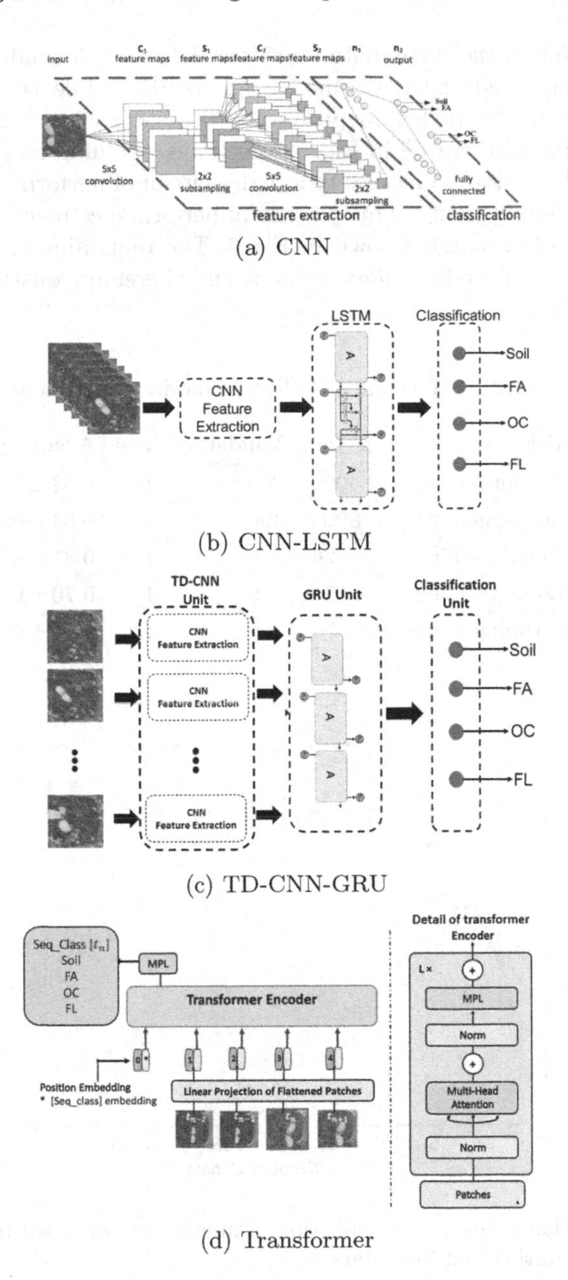

(a) CNN

(b) CNN-LSTM

(c) TD-CNN-GRU

(d) Transformer

Fig. 4. Neural networks architecture tested. Panel (a) Optimized CNN proposed in [15], (b) optimized CNN-LSTM model proposed in [15], (c) optimized TD-CNN-GRU proposed here, (d) transformer adapted from [3].

presented in Sect. 2.2. As visible in Table 2, this simple simulation brings a significant increase of 10% to the overall accuracy on the CNN model. Fine tuning the model trained on these simulated outdoor data with a small amount of real outdoor data improved the performance up to 91% while the model trained on the same amount of real data produced 70% accuracy on the CNN model. Interestingly, as demonstrated in Fig. 5, fine tuning training on data augmented indoor data with shadow converges to a high plateau of performance with a very small number of input plants. This plateau of performance reached with 7 plants produces a confusion matrix shown in Fig. 6. The remaining errors are limited to adjacent classes of seedling development and therefore constitute reasonable errors.

Table 2. Performance of CNN in outdoor conditions.

Models	Train	Validation	Test	Accuracy
Brut transfer	400	200	4	0.53 ± 0.02
Data augmentation	800	400	4	0.64 ± 0.10
Outdoor training	26	6	4	0.81 ± 0.02
Outdoor training	7	6	4	0.70 ± 0.03
Fine tuning training	7	6	4	0.91 ± 0.02

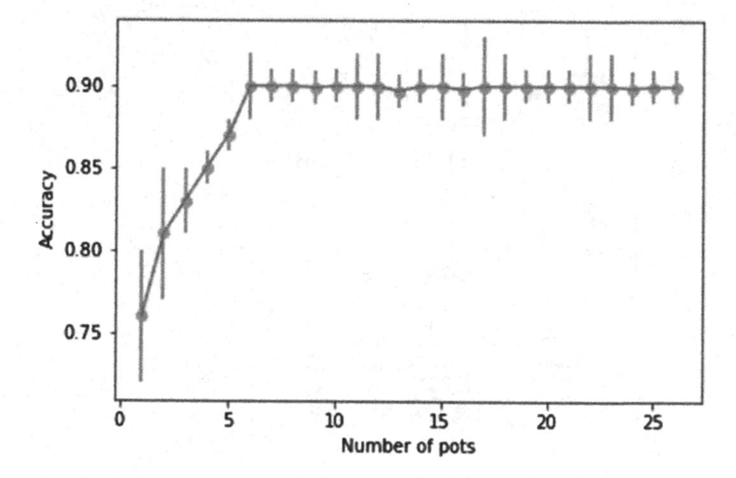

Fig. 5. Classification accuracy as a function of number of pots used in train database after data augmentation and fine tuning.

Similar experiments have been carried with the TD-CNN-GRU model as provided in Table 3 and with the Transformer in Table 4. Indoor classification performances with these spatio-temporal methods were better than the spatial CNN. However, they appear to drop when applied to outdoor data and become less interesting than the pure spatial CNN approach. The data augmentation

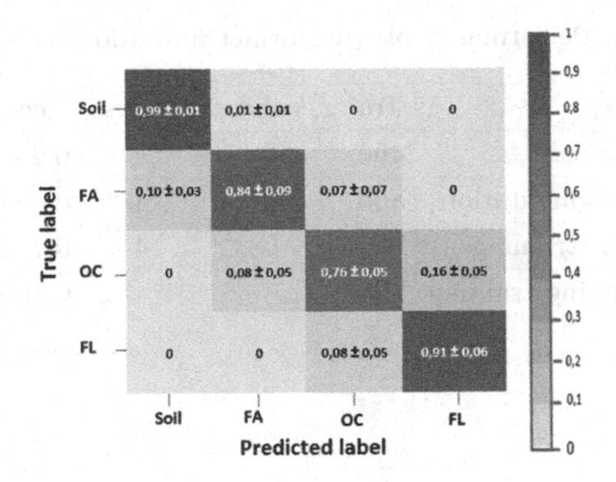

Fig. 6. Confusion matrix of CNN after data augmentation and fine tuning training model using seven pots.

approach with the proposed shadow generator is improving the performance of the TD-CNN GRU and the Transformer by comparison with a direct brut transfer. Yet, they perform in the end with this data augmentation at the same level as if they had been trained fully outdoor.

Several parameters could influence the temporal information from indoor to outdoor. Despite similar speed of the seedling development (approximately 72 h for the whole process on average) for indoor and outdoor conditions, the difference of growing conditions may have influenced the kinetics to pass from one developmental stage to another. Therefore a systematic analysis of the statistics to pass from the developmental stage to another could be interesting to carry out. However, data augmentation with shadow systematically improved all tested methods. This demonstrates that the presence of these shadows is a critical limitation when moving from indoor to outdoor.

Table 3. Performance of TD-CNN GRU in outdoor conditions.

Models	Train	Validation	Test	Accuracy
Brut transfer	400	200	4	0.32 ± 0.04
Data augmentation	800	400	4	0.59 ± 0.04
Outdoor training	26	6	4	0.72 ± 0.04
Fine tuning training	26	6	4	0.74 ± 0.02

Table 4. Performance of transformer in outdoor conditions.

Models	Train	Validation	Test	Accuracy
Brut transfer	400	200	4	0.23 ± 0.03
Data augmentation	800	400	4	0.56 ± 0.04
Outdoor training	26	6	4	0.74 ± 0.03
Fine tuning training	26	6	4	0.76 ± 0.02

4 Conclusion

In this communication, we have investigated the possibility of transfer of knowledge from indoor to outdoor conditions in a plant science application. We have considered the automatic detection of early stages of seedling development to this purpose. While in controlled conditions, time dependence was found to bring additional information, we found that the presence of shadows in outdoor conditions are destroying this information. However, we have demonstrated that transfer of knowledge from indoor was yet possible via the simulation of shadows to be applied to indoor images. We have demonstrated the interest to train on such simulated data and fine tune on limited amount of real data. The proposed approach is of interest in plant science since outdoor conditions are of importance for agricultural practice while indoor conditions have received considerable attention via the development of phenotyping platforms.

The outdoor noise considered here was limited to shadow. However, other sources of noise could also be included to extend the result of this study. This includes for instance the presence of wind causing motion blur which could also easily be simulated with data augmentation following the approach presented in this study.

Acknowledgment. Authors acknowledges Nicolas Mascher from GEVES, France for help in the outdoor data acquisition and plant preparation. This work received funding from H2020 EU project INVITE.

This work was supported by PHENOTIC platform node of the french research infrastructure on plant phenotyping PHENOME-EMPHASIS.

References

1. Cho, K., Van Merriënboer, B., Bahdanau, D., Bengio, Y.: On the properties of neural machine translation: encoder-decoder approaches. arXiv preprint arXiv:1409.1259 (2014)
2. Ding, B., Long, C., Zhang, L., Xiao, C.: ARGAN: attentive recurrent generative adversarial network for shadow detection and removal. In: Proceedings of the IEEE/CVF International Conference on Computer Vision, pp. 10213–10222 (2019)
3. Dosovitskiy, A., et al.: An image is worth 16 × 16 words: transformers for image recognition at scale. arXiv preprint arXiv:2010.11929 (2020)

4. Douarre, C., Crispim-Junior, C.F., Gelibert, A., Tougne, L., Rousseau, D.: Novel data augmentation strategies to boost supervised segmentation of plant disease. Comput. Electron. Agric. **165**, 104967 (2019)
5. Douarre, C., Schielein, R., Frindel, C., Gerth, S., Rousseau, D.: Transfer learning from synthetic data applied to soil-root segmentation in x-ray tomography images. J. Imaging **4**(5), 65 (2018)
6. Garnot, V.S.F., Landrieu, L., Giordano, S., Chehata, N.: Satellite image time series classification with pixel-set encoders and temporal self-attention. In: Proceedings of the IEEE/CVF Conference on Computer Vision and Pattern Recognition, pp. 12325–12334 (2020)
7. Goodman, J.W.: Speckle Phenomena in Optics: Theory and Applications. Roberts and Company Publishers (2007)
8. Le, H., Samaras, D.: Shadow removal via shadow image decomposition. In: Proceedings of the IEEE/CVF International Conference on Computer Vision, pp. 8578–8587 (2019)
9. LeCun, Y., Bengio, Y., Hinton, G.: Deep learning. Nature **521**(7553), 436–444 (2015)
10. Li, Z., Guo, R., Li, M., Chen, Y., Li, G.: A review of computer vision technologies for plant phenotyping. Comput. Electron. Agric. **176**, 105672 (2020)
11. Lobet, G., Draye, X., Périlleux, C.: An online database for plant image analysis software tools. Plant Meth. **9**(1), 1–8 (2013)
12. Patrício, D.I., Rieder, R.: Computer vision and artificial intelligence in precision agriculture for grain crops: a systematic review. Comput. Electron. Agric. **153**, 69–81 (2018)
13. Qu, L., Tian, J., He, S., Tang, Y., Lau, R.W.: DeshadowNet: a multi-context embedding deep network for shadow removal. In: Proceedings of the IEEE Conference on Computer Vision and Pattern Recognition, pp. 4067–4075 (2017)
14. Samiei, S., Rasti, P., Richard, P., Galopin, G., Rousseau, D.: Toward joint acquisition-annotation of images with egocentric devices for a lower-cost machine learning application to apple detection. Sensors **20**(15), 4173 (2020)
15. Samiei, S., Rasti, P., Vu, J.L., Buitink, J., Rousseau, D.: Deep learning-based detection of seedling development. Plant Meth. **16**(1), 1–11 (2020)
16. Sapoukhina, N., Samiei, S., Rasti, P., Rousseau, D.: Data augmentation from RGB to chlorophyll fluorescence imaging application to leaf segmentation of *Arabidopsis thaliana* from top view images. In: Proceedings of the IEEE/CVF Conference on Computer Vision and Pattern Recognition Workshops, pp. 4321–4328 (2019)
17. Vaswani, A., et al.: Attention is all you need. arXiv preprint arXiv:1706.03762 (2017)
18. Yin, W., Kann, K., Yu, M., Schütze, H.: Comparative study of CNN and RNN for natural language processing. arXiv preprint arXiv:1702.01923 (2017)
19. Yuan, Y., Lin, L.: Self-supervised pre-training of transformers for satellite image time series classification. IEEE J. Sel. Top. Appl. Earth Obs. Remote Sens. **14**, 474–487 (2020)
20. Zhou, K., Wang, W., Hu, T., Deng, K.: Time series forecasting and classification models based on recurrent with attention mechanism and generative adversarial networks. Sensors **20**(24), 7211 (2020)

Banana Ripening Classification Using Computer Vision: Preliminary Results

Matheus T. Araujo[1], Miguel W. de V. Santos[1], Flávio F. Feliciano[2], Pedro B. Costa[3], and Fabiana R. Leta[1](✉)

[1] Fluminense Federal University, R. Passo da Pátria 156, Niterói, RJ, Brazil
{araujo_matheus,miguelwenzel,fabianaleta}@id.uff.br
[2] Fluminense Federal Institut, Cabo Frio, RJ, Brazil
[3] Federal University of Minas Gerais, Av. Antonio Carlos 6627, Belo Horizonte, MG, Brazil

Abstract. Color is one of the main features for the evaluation of fruit ripeness. Part of the quality control is based on the ripening status, where fruits should be placed with others with the same ripening phase. However, this color analysis can be done automatically using computer vision techniques. This work presents an evaluation of the ripeness status of bananas, based on the analysis of the texture features obtained with a digital camera. The novelty lies in assessing which parameters best define the maturation stage to automate the classification process with accuracy. The results aim to determine which texture feature can provide better results for the automatic classification of the bananas ripening phase, which is an important challenge to food industry for export.

Keywords: Computer vision · Banana ripening · Texture feature

1 Introduction

For many years, food selection processes were carried out, necessarily, by a human operator, only using visual inspection. It is one of the oldest existing techniques and it is widely used for the evaluation of product quality conditions due to the easiest execution and because it does not require special equipment. However, with technological advances, food industries and large producers, aiming mostly at exporting products, are demanding advances in the selection process in other to increasing product analysis.

One of the biggest challenges for image inspection systems consists in joining the quality of data acquisition with the reduction of costs and losses during the process, considering the reliability and accuracy provided (Gomes and Leta 2012). The development of an inspection system should consider that it can be performed in the same way and with the same quality responses anywhere it is applied.

In Brazil, banana production in 2018 was 6,752,171 tons, accounting for approximately 2% of this year's agricultural production. Due to this advance and the high demand for banana exports in Brazil, automated inspections become an important solution for quality control involving the evaluation of color, texture and shape, which are the characteristics observed by consumers when purchasing the product and in cultivation

G. Rozinaj and R. Vargic (Eds.): IWSSIP 2021, CCIS 1527, pp. 132–139, 2022.
https://doi.org/10.1007/978-3-030-96878-6_12

monitoring (Gomes and Leta 2012; Jaffery and Dubey 2016; Hou et al. 2015; Mazen and Nashat 2019). In the international market, bananas are one of the most commercialized fresh fruits. For its commercialization fruits are classified according to the stage of ripening, mainly on the basis of peel color (Gomes et al. 2014).

The banana peel color is not homogeneous during ripening. This characteristic can be evaluated from the perspective of texture analysis. Adebayo et al. (2016) affirm that ripening in fruit is correlated with alterations in the fruit texture which are more pronounced in climacteric fruits such as banana. In this context, this paper presents an evaluation of the ripeness state of bananas, essential for the commercialization and exportation of this product, based on a peel texture analysis.

Texture analyses methods have been used in many applications, especially in recognizing the material corrosion (Xia et al. 2020; Da Silva et al. 2015; Medeiros et al. 2010; Ahuja and Shukla 2017). The presented methodology is based on a similar approach used to analyze the changes in steel surface appearance, considering its corrosion degradation. Feliciano et al. (2015) proposed the use of texture analysis for nondestructive surface corrosion monitoring. They evaluated six textural characteristics and the obtained result shows that the technique was feasible as a new method to check the surface corrosion state.

2 Texture Analysis

Texture is characterized by the image repetition (textel) in a given region. The textel can be repeated on the image with variations in size, intensity, color and orientation and still contain noise. The texture analysis goal is to identify the neighbourhood of these similar elements that characterize the connectivity, density, and homogeneity.

Some techniques can be found in the literature to identify texture features. Most of them are based on greyscale pixels. A widely used texture feature is entropy. Entropy is a concept used in thermodynamics to measure the degree of organization of a system. However, this concept was introduced to other areas of knowledge. As information theory and, consequently, in pattern recognition. Therefore, entropy is a statistical measurement of randomness of intensity distribution.

There are techniques that take into account the pixels' spatial relationship (frequency $p(i,j)$ at which a pixel with a grey level i and another with grey level j occur in an image separated by a distance d). From relationship matrices, it is possible to calculate numerical values called descriptors that provide information about the original image; among them are probability, differences moments, energy, variance, correlation, homogeneity, and others. To name the most common: Contrast is the measure of intensity contrast between a pixel and its neighbour, correlation is the statistical measure of how correlated a pixel is to its neighbour over the whole image and homogeneity is the measure of closeness of the distribution of elements in relationship matrix.

The most appropriate texture parameter for each application should be defined based on the context because despite the existence of several methods, none are able to effectively target all types of texture.

3 Materials and Methods

In this paper we use the visual classification standards considering environment conditions determined by the Brazilian Program for the Modernization of Horticulture (PBMH and PIF 2006). This classification is based on the observation of the ripening process in the banana's peel, where the color changes is intense. Changes observed from the green due the high chlorophyll rate, to the yellow that appears in consequence of the chlorophyll degradation along the ripening process. The favorable environmental conditions for the storage of bananas are temperature between 14 °C to 24 °C and relative humidity between 80 and 90%. The maturation speed is proportional to the temperature, that is, the higher the temperature, the faster the maturation because of the dehydration. Maintaining the humidity in the mentioned range causes the longevity of the fruit.

In the experiment six Brazilian 'Prata' banana (*Musa sapientum AAB*) were exposed to the given condition in a selected environment. The details of the image acquisition, the condition and the environment will be described in the following items.

3.1 Computer Vision System

Image Acquisition

Image acquisition was carried out using a led ring with light diffuser, 10 W power and 5500 K color temperature. A smartphone camera with following setup: the automatic ISO, ranging between 100 and 140, auto focus with fixed camera aperture and JPG format with 7.9 MP.

Figure 1 shows the light camara prototype.

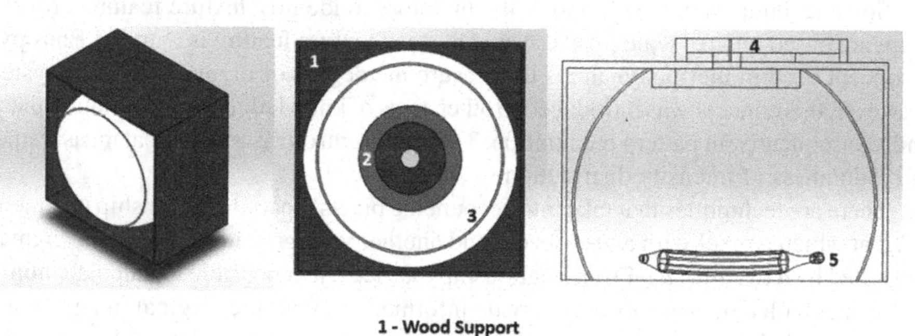

Fig. 1. Light camara prototype.

Samples
The samples are the banana units detached from the bunch of Brazilian 'Prata' type, purchased in a street market in their stage 3, according to Von Loesecke's scale (Gomes et al. 2014). The samples don't have any additives on their peel and they were cleaned.

Environment
In this experiment the bananas were exposed to an ordinary environment condition in Brazil that is characterized by being a small room, but with good air circulation, which avoids large variations in temperature and humidity, and protected from possible rain and direct solar radiation on the fruits, being beneficial for a minimum control of these variables since it is not possible to perform in a properly controlled laboratory environment due to the restrictions imposed to contain the new COVID-19 pandemic. In this environment the bananas were exposed to the ambient air without any direct protection, aiming to maintain the common ripening process of the fruit.

The banana goes through the ripening process varying according to the temperature conditions and relative air humidity to which they are submitted, thus, considering the conditions and the ripening point at the moment of purchase of the bananas to be analyzed, the time of analysis was 80 h, 3 days and 8h, with photos taken every 8h totaling, counting the photo from the initial point 0h, 11 photos per banana.

3.2 Image Processing

In the tests performed the registrations were made in such a way that the distance between camera and object, the camera angle and the lighting were the same for all bananas, thus ensuring that the algorithms run the same way in all registered images and reducing the possibilities of undesired variations between photos of the same banana.

Regarding the dimensions, the images obtained at a distance of 23 cm from the base have dimensions 1940 $\times$ 4096 pixels. Each image, due to the reduction of brightness and shadows, is used completely for the beginning of the processing, that is, without any previous cut.

There are 6 bananas for each condition, totaling 24 bananas in all. As mentioned in item 3.1, each banana was photographed in 11 moments throughout the test, which generates 66 images in the interval of 80 h. Figure 2 shows an example of an image sequence of one of the 6 bananas.

For the ripening process analysis, all images were processed using the methods described in 2.2, i.e., processed from the grayscale version in order to perform the segmentation and extract the values of the following texture characteristics: percent cover; entropy; contrast; color-pixel ratio; energy; homogeneity.

The segmentation was performed in two stages, the first of them performing the separation between object and background through automatic thresholding, obtained through the image histogram, and, after this first stage, it is processed a morphological filter to fulfil distortions due the segmentation in the banana peel.

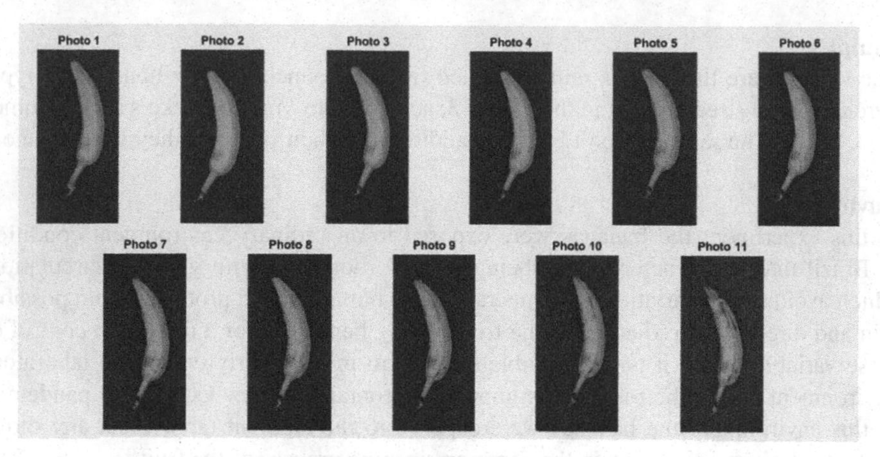

Fig. 2. Example of a sequence of images taken from one of the 6 bananas analyzed (Color figure online)

4 Results

The aim of this step is to identify which of the features extracted from bananas can be considered as the best candidate for a future stage to be applied in the automatic ripening classification.

Thus, the technique must present a significant discrepancy when the ripenning class changes and the numerical differences are not significant while the fruit remains in the same stage of ripeness.

The first analysis carried out was a visual analysis of the graphs of the extracted features, as a way of observing the behaviors of the characteristics over the 80 h of samples monitoring.

Through this first analysis, the characteristics Contrast and Homogeneity exhibited constant characteristics in thc graph, either of growth or decrease over time during the analysis.

4.1 ANOVA

The second step was the application of statistical techniques in order to identify the numerical variations in the chosen characteristics and when these variations are significant enough to identify a change in class during the experiment.

The data were organized to perform a one-factor test. Where the factor was the time, divided into 11 levels, and for each level, 6 replicates of the experiment were performed, as shown in Table 2.

Table 2. Results from the homogeneity feature

	Homogeneity										
Samples	0 h	8 h	16 h	24 h	32 h	40 h	48 h	56 h	64 h	72 h	80 h
B1	0,894	0,882	0,871	0,868	0,848	0,850	0,826	0,815	0,798	0,777	0,774
B2	0,888	0,877	0,867	0,862	0,849	0,843	0,821	0,811	0,799	0,779	0,753
B3	0,883	0,880	0,872	0,867	0,850	0,851	0,839	0,823	0,809	0,796	0,773
B4	0,906	0,900	0,894	0,886	0,866	0,858	0,840	0,852	0,835	0,822	0,798
B5	0,896	0,893	0,888	0,884	0,877	0,866	0,854	0,825	0,806	0,802	0,798
B6	0,894	0,888	0,876	0,869	0,851	0,843	0,821	0,808	0,807	0,810	0,791

For the test, the hypothesis of whether or not a term τ_i exists that causes changes between the levels of the factor.

$$H_0 : \tau_0 = \tau_1 = \tau_2 = \cdots = \tau_a$$
$$H_1 : \tau_i \neq 0 \textit{ at least one } i \tag{1}$$

Applying the F test, the p value obtained was $p = 1.4 \times 10^{-19}$ for homogeneity and $p = 1.1 \times 10^{-13}$ for contrast feature. Both values provide significant evidence of rejection of the null hypothesis of variations being equal to zero between the 80 h in which the samples were captured.

In order to try to identify when the variations are significant, the Minimum Significant Difference method was used. The method provides, based on the significance of 0.05, limits in which the differences between the averages of the levels can be considered statistically equal or not.

With the application of the MDS, 2 class variations were identified for the contrast and 4 using homogeneity. When comparing the visual analyzes, the method does not present significant results, since the visual analysis identified 5 or 6 classes, depending on the operator.

4.2 Linear Regression

To improve the numerical analysis and find a method that is able to find the differences between the classes more closely than the differences found by the visual ripening analysis, a linear regression was used.

For this, linear equations were adjusted for the averages of the 6 samples for each of the 11 evaluated points.

After the curve adjustments, confidence intervals were calculated for each of the points obtained in the experiments, as shown in figure X.

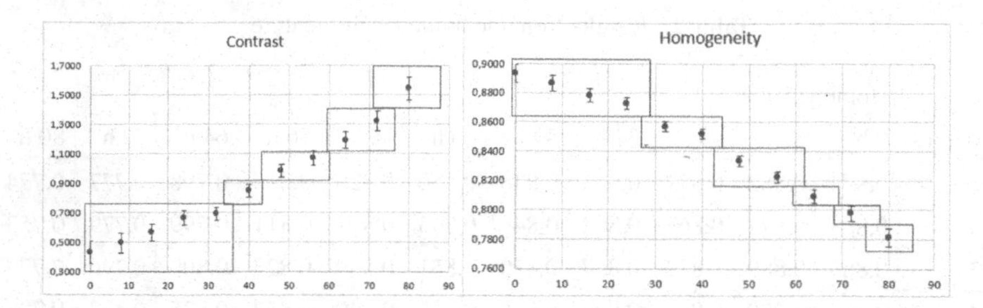

From the confidence intervals for each sampling performed, it is possible to infer that the characteristic of homogeneity identified significant statistical variations allowing the identification of 6 different classes, while the contrast characteristic identified 5 different groups.

When comparing the visual evaluations performed by two operators, the results are quite significant, since one of the operators identified five different classes during the 80 h of the experiment and a second operator identified six classes.

5 Conclusions

The present work aimed to evaluate the use of characteristics, typical of texture analysis used in computer vision, to determine the status of banana ripening.

Among the characteristics analyzed, contrast and homogeneity shown a linear behavior throughout the 80 h of sampling.

In a second stage, statistical techniques were used in order to evaluate if during the 80 h of sampling could be identify significantly differences between texture values.

The techniques of homogeneity and contrast, observed by a regression analysis showed that it was possible to identify a similar number of classes to that observed in a visual analysis performed by human operators.

In a next step, new tests will be carried out to validate the automatic classification of the banana ripening phase using the procedure and the proposed characteristics.

References

Adebayo, S.E., Hashim, N., Abdan, K., Hanafi, M., Mollazade, K.: Prediction of quality attributes and ripeness classification of bananas using optical properties. Sci. Hortic. **212**, 171–182 (2016)

Ahuja, S.K., Shukla, M.K.: A survey of computer vision based corrosion detection approaches. In: Satapathy, S.C., Joshi, A. (eds.) ICTIS 2017. SIST, vol. 84, pp. 55–63. Springer, Cham (2018). https://doi.org/10.1007/978-3-319-63645-0_6

Da Silva, N.R., et al.: Improved texture image classification through the use of a corrosion-inspired cellular automaton. Neurocomputing **149**, 1560–1572 (2015)

Feliciano, F.F., Leta, F.R., Mainier, F.B.: Texture digital analysis for corrosion monitoring. Corros. Sci. **93**, 138–147 (2015)

Gomes, J.F.S., Leta, F.R.: Applications of computer vision techniques in the agriculture and food industry: a review. Eur. Food Res. Technol. **235**, 989–1000 (2012)

Gomes, J.F.S., Vieira, R.R., de Oliveira, I.A.A., Leta, F.R.: Influence of illumination on the characterization of banana ripening. J. Food Eng. **120**, 215–222 (2014)

Hou, J.C., Hu, Y.H., Hou, L.X., Guo, K.Q., Satake, T.: Classification of ripening stages of bananas based on support vector machine. Int. J. Agric. Biol. Eng. **8**(6), 99–103 (2015)

Jaffery, Z.A., Dubey, A.K.: Scope and prospects of non-invasive visual inspection systems for industrial applications. Indian J. Sci. Technol. **9**(4), 15 (2016)

Lima, M.B., Silva, S. de O.E., Ferreira, C.F.: Banana: o produtor pergunta, a Embrapa responde, 2 ed. rev. e ampl. Embrapa, Brasília, DF (2012)

Mazen, F.M.A., Nashat, A.A.: Ripeness classification of bananas using an artificial neural network. Arab. J. Sci. Eng. **44**(8), 6901–6910 (2019). https://doi.org/10.1007/s13369-018-03695-5

Medeiros, F.N.S., et al.: On the evaluation of texture and color features for nondestructive corrosion detection. EURASIP J. Adv. Signal Proc. **2010**, 1–7 (2010)

Xia, D., et al.: Review-material degradation assessed by digital image processing: fundamentals, progresses, and challenges. J. Mater. Sci. Technol. **53**, 146–162 (2020)

Energy Reconstruction Techniques in TileCal Under High Pile-Up Conditions

Guilherme Inácio Gonçalves[1,2(✉)] on behalf of the ATLAS Tile Calorimeter Group

[1] Federal University of Rio de Janeiro, Rio de Janeiro, RJ, Brazil
ginaciog@cern.ch

[2] Computational Modeling Graduate Program, Rio de Janeiro State University, Nova Friburgo, Brazil

Abstract. Particle colliders are machines built to probe fundamental questions in physics. The properties of the produced particles are measured by complex experiments which use a wide variety of devices, such as the calorimeter system. In the ATLAS experiment at LHC, the Tile Calorimeter (TileCal) comprises about 10,000 readout channels that amplify, shape, and sample each signal every 25 ns. Since LHC collisions occur every 25 ns, and due to the increase of the luminosity level, signals from adjacent collisions may be read out within the same TileCal readout window, deforming the expected signal, and degrading the energy estimation efficiency. Therefore, this work compares the performance of the currently available methods for TileCal energy estimation using LHC collision data. Different pile-up conditions are considered. The results show that the performance in terms of the uncertainty of the energy estimation can be improved up to 35% in high-occupancy readout channels.

Keywords: Optimal filter · Signal estimation · Wiener filtering · Pile-up · High-energy calorimetry.

1 Introduction

The electronics for high-energy physics experiments deal with enormous technological challenges as a large amount of data needs to be processed within a short time. The calorimeter systems of such complex experiments play an important role as they are used to measure the energy of incident particles. The information provided by the calorimeter systems is used for event reconstruction and particle identification [23].

Typically, a calorimeter is segmented into readout cells (tens of thousands in modern calorimeters), providing spatial resolution to the detector. The readout signals are processed by a pulse-shaping electronic circuit, which gives the pulse of a well-defined shape with an amplitude proportional to the particle energy [15]. Thus, the problem of energy estimation can be stated as determining the amplitude of the pulse produced by the calorimeter readout channel.

G. Rozinaj and R. Vargic (Eds.): IWSSIP 2021, CCIS 1527, pp. 140–151, 2022.
https://doi.org/10.1007/978-3-030-96878-6_13

Energy is the most important information in high-energy calorimetry systems, being a fundamental property for the reconstruction of events and validation of models for physical phenomena.

The commonly employed mathematical methods for energy estimation formulate the problem as estimating the amplitude of a pulse immersed in additive noise, where the identification of channels with relevant information is performed through an energy threshold at the output of the estimator. Usually, when modeling the problem, these approaches consider a fixed pulse shape and additive Gaussian noise [1,11].

However, modern colliders, such as the LHC [10], operate with high event rates and high luminosity levels. The luminosity is defined as a proportional factor between the number of events per second and the interaction cross section, having the unit of cm^2s^{-1} [13]. In such conditions, the particle density in the beam cross-section is increased so that many interactions could occur at each collision point, generating more signals in the detectors, such as the calorimeters [16]. As a result, the energy estimation problem becomes more complex because of that can be observed where two or more pulses are acquired within the same readout window. To mitigate the, new methods based on the deconvolution of superimposed signals have been proposed [3,6]. Additionally, an approach based on Wiener filtering has also been tested for severe conditions [19].

The ATLAS (*The Toroidal LHC AparatuS*) experiment [4] covers a wide spectrum of physics of interest at the LHC, and the information from its calorimeter system is important for the complex trigger system which selects only the relevant information from the collisions to be stored. Future LHC upgrades are planned to increase the collision energy and the luminosity level. The increase in luminosity raises the number of proton-proton interactions per collision, producing more data and increasing the probability of observing events of interest. However, in the calorimeter system, signals from neighboring events that occurred in different time may be read out causing the effect which degrades the performance of typical energy estimation methods. In the Tile Calorimeter (TileCal) [5] of the ATLAS, three energy estimation methods address the in different ways. Therefore, this work evaluates the performance of these methods in different conditions using collision data acquired during 2018 data-taking.

In the next section, the TileCal is briefly introduced. In Sect. 3 the methods used in TileCal for energy estimation are described. The results describing the performance analysis of the methods using real proton-proton collision data acquired in 2018 in ATLAS are presented in Sect. 4. Finally, in Sect. 5 conclusions are derived.

2 The ATLAS Tile Calorimeter

The ATLAS calorimetry system is sectioned according to the interaction nature: electromagnetic (electrons and photons) and hadronic (e.g. protons and neutrons). The Tile Calorimeter (TileCal) is the central hadronic calorimeter of ATLAS and provides accurate measurements of jet energy, and assists in the missing transverse moment calculation and muon detection.

TileCal is a sampling calorimeter that uses scintillating plastic plates (or tiles) as the active material, interspersed with steel layers as the absorbent material, according to the illustration of a module depicted in Fig. 1. Both sides of each tile are connected to a specific type of optical fiber, called WLS (*Wavelength Shifting*), composed of a photo-fluorescent material that absorbs a high-frequency photon and emits multiple low-frequency photons. Multiple tiles are grouped into cells and each cell is connected to two photo-multipliers (PMTs). TileCal is composed of four partitions (EBC, LBC, LBA, and EBA). Each partition is divided into 64 modules with 32 to 48 readout channels per module, producing approximately 10.000 readout signals per collision.

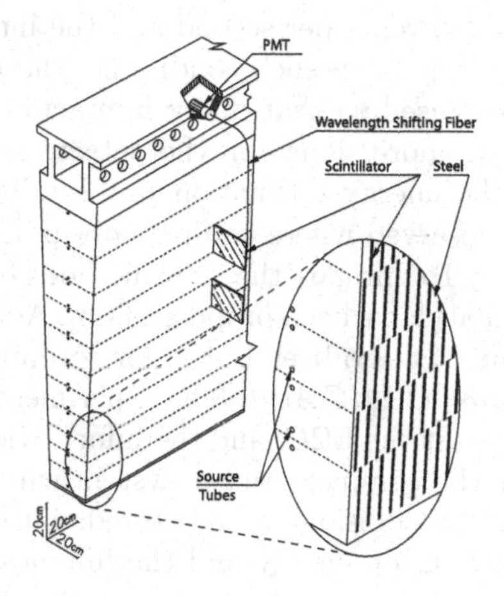

Fig. 1. Schematic diagram of a single TileCal module (extracted from [7]).

When a hadronic particle passes through the calorimeter, it generates a shower of secondary particles and loses energy by the interaction with the passive material (steel as an absorbent material) and also interacts with the active material (scintillating tiles), emitting light. This light is conducted by optical fibers and read out by PMTs, which generate an electrical pulse in response to the light signal. The pulse generated by the PMT is conditioned by a conformation circuit (*shaper*), which provides a pulse with a known shape and amplitude proportional to the deposited energy [2]. This analog pulse is digitized by an Analog to Digital Converter (ADC) with a sampling frequency of 40 MHz. A window with seven time samples is available to extract the parameters from the readout pulse.

The deposited energy in each cell of the calorimeter can be calculated by correctly estimating the pulse amplitude, which is an approach commonly used in modern calorimeters. Physically, only cells located along the shower development should contain energy deposits and are selected for energy reconstruction [21].

The average number of interactions per collision $<\mu>$ is used to represent how occupied a given readout cell is at each collision. Currently, the LHC operates with <μ> approximately equal to 40, and this value should increase for *Run 3* [22]. As a consequence of the increase in luminosity, the probability of the occurrence of signals within the same TileCal readout channel also increases. When signals from adjacent events are acquired within a same readout window, the effect is observed as illustrated by Fig. 2.

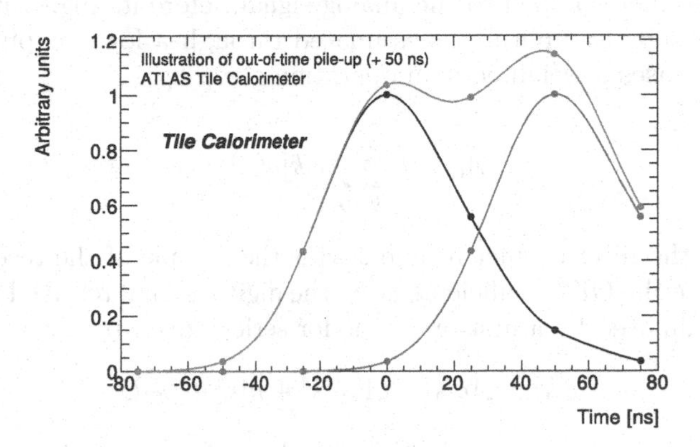

Fig. 2. An illustration of the pile-up effect. The black pulse is the signal of interest and the red one is the time-shifted signal. The resulting received pulse is in magenta (extracted from [14]). (Color figure online)

3 Energy Estimation in TileCal

The problem of the energy reconstruction in calorimeters is commonly addressed as a parameter estimation problem. Currently, TileCal has three methods available for energy reconstruction of signals from collisions: the Optimal Filter, the Constrained Optimal Filter (COF), and the Wiener Filter. The methods address the in different ways as described below.

3.1 The Optimal Filter

The Optimal Filter (OF) method aims to estimate the amplitude of an input signal, which is proportional to the energy. This estimator is designed to minimize the estimation variance using the knowledge of the pulse shape (output of the shaper circuit). The main noise source in ATLAS calorimeters readout channels is the electronics, which can be modeled by a Gaussian distribution. Under this constraint, the filter operates optimally which makes this method widely used in ATLAS [9,18].

The implemented version of this estimator in TileCal is called OF2 and has been in operation since 2014, being used for online and offline energy reconstruction [11]. This method is also used in other calorimeters in ATLAS, such as the Liquid Argon Calorimeter (LAr) [18].

In the OF2 method, the digital sample $x[k]$ at time k can be modeled by

$$x[k] = Ag[k-\tau] + n[k] + ped\,, \tag{1}$$

where A is the pulse amplitude, $g[k]$ the reference pulse values (shaper), $n[k]$ the additive noise, τ the phase shift of the signal, and *ped* the pedestal or baseline, a constant parameter added to the analog signal before its digitization.

The input signal amplitude is estimated through a low computational cost process, which uses a weighted sum operation given by

$$\hat{A}_{OF} = \sum_{k=0}^{N-1} x[k]w[k], \tag{2}$$

where $w[k]$ is the filter coefficient and $x[k]$ is the sample of the received signal.

To calculate the OF2 coefficients $w[k]$, the digital signal received from TileCal can be approximated by a first-order Taylor series, given by

$$x[k] = Ag[k] - A\tau\dot{g}[k] + n[k] + ped, \tag{3}$$

where $\dot{g}[k]$ represents the derivative of the reference pulse g, and $k = 0, 1, 2, \ldots, N-1$.

To guarantee an unbiased estimator, the absence of estimation bias, the expected value of $\hat{A}_{OF}$ is required to equal A. Therefore, replacing Eq. (3) in (2) and considering that the average noise is zero ($\mathbb{E}\{n[k]\} = 0$), where $\mathbb{E}$ represents the expectation operator, the expected amplitude value becomes

$$\mathbb{E}\{\hat{A}_{OF}\} = \sum_{k=0}^{N-1} (Aw[k]g[k] - A\tau w[k]\dot{g}[k] + w[k]ped), \tag{4}$$

and

$$\mathbb{E}\{\hat{A}_{OF}\} = A\,. \tag{5}$$

For the estimator to be independent of the pedestal and the phase, the following restrictions are established

$$\sum_{k=0}^{N-1} w[k]g[k] = 1, \sum_{k=0}^{N-1} w[k]\dot{g}[k] = 0, \text{and} \sum_{k=0}^{N-1} w[k] = 0. \tag{6}$$

The first restriction guarantees an unbiased estimator, while the second and third restrictions guarantee, respectively, that the estimator is immune to phase and pedestal fluctuations.

The estimator's variance is given by

$$\begin{aligned}\mathbb{E}\{(\hat{A}_{OF} - A)^2\} &= \sum_{k=0}^{N-1}\sum_{j=0}^{N-1} w[k]w[j]C[k,j] \\ &= \mathbf{w}^T\mathbf{C}\mathbf{w},\end{aligned} \tag{7}$$

where $\mathbf{w}$ is the weights vector of the estimator and $\mathbf{C}$ the noise covariance matrix.

To determine the weights $\mathbf{w}$, it is necessary to minimize the expression of the estimator's variance using the Lagrange multiplier method. The solution of this system results in the set of weights $w[k]$ of the OF2 estimator that currently operate online and offline in TileCal. It is worth mentioning that currently the noise covariance matrix $\mathbf{C}$ is approximated by the identity matrix, which does not take into account the effect.

3.2 The COF Method

Another algorithm also available for offline reconstruction in TileCal is the COF method (*Constrained Optimal Filter*). COF computes a linear transformation that recovers the amplitude of superimposed signals for a given readout window. Hence, the central pulse, assigned to the collision of interest, can be dissociated and reconstructed [3].

In this respect, the COF method models the energy deposition in a given calorimeter cell as a Kronecker delta function [20], which produces an output corresponding to the TileCal reference pulse. Thus, considering a set of energy depositions $a[k]$, the received signal can be modeled as

$$x[k] = \sum_{i}(g[i]a[n-i]) + n[k]. \tag{8}$$

In this way, estimating the deposited energy in a given calorimeter cell implies deconvolution of the sequence $x[k]$ of the impulse response $g[k]$. Applying a similar procedure as for the OF method, considering the vector of $x[k]$ (represented as $\mathbf{x}$) time samples, the j set of amplitudes $\hat{\mathbf{a}}_j$ can be given by

$$\hat{\mathbf{a}}_j = \mathbf{U}_j^T\mathbf{x}, \tag{9}$$

where

$$\mathbf{U}_j = \mathbf{C}_j^{-1}\mathbf{G}_j(\mathbf{G}_j^T\mathbf{C}_j^{-1}\mathbf{G}_j)^{-1}. \tag{10}$$

The $\mathbf{G}_j$ parameter corresponds to the matrix of shifted versions of TileCal reference pulse, where j is the number of collisions within the calorimeter readout window, and $\mathbf{C}_j$ is the noise covariance matrix. When $j = N$, the number of collisions is equal to the size of the readout window, thus the estimator will take the form

$$\hat{\mathbf{a}} = \mathbf{G}_j^{-1}\mathbf{x}. \tag{11}$$

It is also worth pointing out that this expression does not depend on the noise covariance matrix $\mathbf{C}$, which is one of the advantages of the COF method over the OF method. Finally, COF method applies a linear cut to select only the amplitudes above a predefined threshold, defined in the filter design. This step aims at re-designing COF through Eq. (10), avoiding estimating signals without information (noise), and improving the amplitude estimates with relevant information. The drawback is that signals outside the readout window are not considered in the COF design.

3.3 The Wiener Filtering

A third recently implemented and validated approach for energy estimation in TileCal readout channels is based on the Wiener filtering. In this method, a digital linear filter $c[0], c[1], \ldots, c[N-1]$ is designed, where the output $y[N]$ provides an estimate of the desired response $d[n]$ (acquired through simulation), given an input signal with N elements $x[0], x[1], \ldots, x[N-1]$. The Wiener filter design aims to minimize the mean square value of the estimation error, which leads to a mathematically more treatable problem. Unlike the OF method, the Wiener filter considers the uncertainties from the signal and the noise in its minimization process. In particular, this criterion based on the mean square error results in a second-order dependence of the cost function on the filter coefficients. Furthermore, the cost function has a distinct global minimum that defines singularly the optimal design of the filter, in the statistical sense [12]. In this approach, the filter output is given by the sum

$$y[n] = \sum_{k=0}^{N-1} c[k]x[n-k], \tag{12}$$

such that the error between the desired value and the estimated value $e[n] = d[n] - y[n]$ is minimized. To optimize the filter design, the criterion of minimizing the mean square error was adopted. For this, the following cost function is defined by

$$\mathbf{J} = \mathbb{E}\{e[n]^2\}. \tag{13}$$

The minimum of the cost function $\mathbf{J}$ in respect to the coefficients $c[k]$ is given by

$$\sum_{i=0}^{N-1} c[i]\mathbb{E}\left\{x[n-k]x[n-i]\right\} = \mathbb{E}\left\{x[n-k]d[n]\right\}, \tag{14}$$

where $k = 0, 1, \ldots, N-1$. From Eq. (14), it can be seen that:

1. The expected value $\mathbb{E}\left\{x[n-k]x[n-i]\right\}$ is the auto-correlation function of the filter input for the $i-k$ lag. This expression can be rewritten as

$$R[i,k] = \frac{1}{N}\sum_{n=0}^{N-1} x[n-k]x[n-i]. \tag{15}$$

2. The expected value $\mathbb{E}\left\{x[n-k]d[n]\right\}$ is the cross-correlation between the filter input and the desired output for the $i-k$ lag. This expression can also be rewritten as

$$p[k] = \frac{1}{N}\sum_{n=0}^{N-1} x[n-k]d[n]. \tag{16}$$

Equations (15) and (16) are known as the Wiener-Hopf equations. It should be stressed that for nongaussian noise (such as pile-up), this method operates in sub-optimal conditions. Substituting these equations in Eq. (14), a linear equation system is obtained as a necessary and sufficient condition to optimize the filter, as follows

$$\sum_{i=0}^{N-1} c[i]R[i,k] = p[k] \quad k = 0, 1, \ldots, N-1. \tag{17}$$

Finally, this equation system can be rewritten in the matrix form and the Wiener filter optimal weights can be expressed by

$$\mathbf{c} = \mathbf{R}^{-1}\mathbf{p}, \tag{18}$$

where $\mathbf{R}$ represents the auto-correlation matrix of the input signals samples (Eq. (15)) and $\mathbf{p}$ represents the cross-correlation matrix between the input signals samples and the desired values for the filter output (Eq. (16)).

It is worth mentioning that Wiener filter results in the optimal filter in the sense of minimizing the mean square error (error dispersion), taking into account the statistics present in the input data (signal plus noise). However, the Wiener filter considers that the noise has a zero mean (average value), which does not correspond to the case of uni-polar signals pile-up noise. To circumvent the noise average problem, an additional coefficient is included in the Wiener filter optimization process. This additional element of constant value equal to 1 is added to each input signal as the last element. In this way, the input signal has $N+1$ elements and the coefficient vector is also increased by one element.

The goal of including this additional element is to cancel the independent component of the signal in the optimization procedure, absorbing the average noise value in order to compensate for its contribution in the amplitude measurement of a given readout window. As a result, the estimation of the $\hat{A}_{FW}$ amplitude of the proposed Wiener filter is given by the sum of the products of the received signal temporal samples and the first N coefficients of $\mathbf{c}$. At the end of the operation, the last coefficient $c[N]$ is added to the result, compensating for the average noise value as shown in Eq. (19).

$$\hat{A}_{FW} = \left(\sum_{i=0}^{N-1} c[i]x[i] \right) + c[N]. \tag{19}$$

4 Results

Real proton-proton collision data acquired in 2018 by the LHC (last data acquisition period) were used for performance evaluation of the described methods [17]. Different levels were tested, with an average number of interactions per collision $<\mu> \approx 30$, $<\mu> \approx 40$, $<\mu> \approx 50$, and $<\mu> \approx 90$. The datasets for each case contain about 1 million events. For $<\mu> \approx 30$ and $<\mu> \approx 50$, the Wiener filter method was designed using a dataset of $<\mu> \approx 40$, in order to profit of the number of signals used to design the methods. For $<\mu> \approx 90$, the dataset was divided equally into two subsets, the development set (used to design the methods), and the test set (used to evaluate their efficiency). Since the OF and COF methods use only the pulse shape information in their models, the same design of each method was used for all $<\mu>$ values.

It is worth mentioning that the data used were obtained from datasets called *ZeroBias Stream*, where only random triggers are used. Therefore, only electronic noise and pile-up information were acquired. In other words, in these events it is not expected to observe any signal of interest in the acquisition window, constituting only noise data. Hence, the mean value and RMS of the energy estimation of these events represent, respectively, the estimation bias and variance associated with each algorithm.

The Wiener filter is designed from a dataset composed of signals of interest immersed in noise, as well as the known amplitude values of the respective signals. Therefore, a pulse simulator was developed and validated to produce a dataset of signals of interest with the known amplitudes values, considering both pulse deformation and phase shift uncertainties. The amplitude follows a uniform distribution in the range of $[0, 1023]$ ADC counts, since TileCal's analog-to-digital converter has 10 bits [2]. Thus, each amplitude value has the same probability of occurrence. Finally, the generated signals were added to the events of the dataset used to derive the Wiener filter coefficients.

4.1 Efficiency Analysis

To analyse the efficiency of the studied filters for severe conditions, the most affected cells in terms of effect in TileCal, called E4 cells, were used. Figs. 3a and b show the reconstructed energy distributions by the Wiener filter, COF, and OF2, considering $<\mu> \approx 50$ and $<\mu> \approx 90$, respectively. It can be seen the histograms for the Wiener filter shows less dispersion followed by the COF method. This result shows that for channels with high pile-up levels the Wiener filter presents a promising performance.

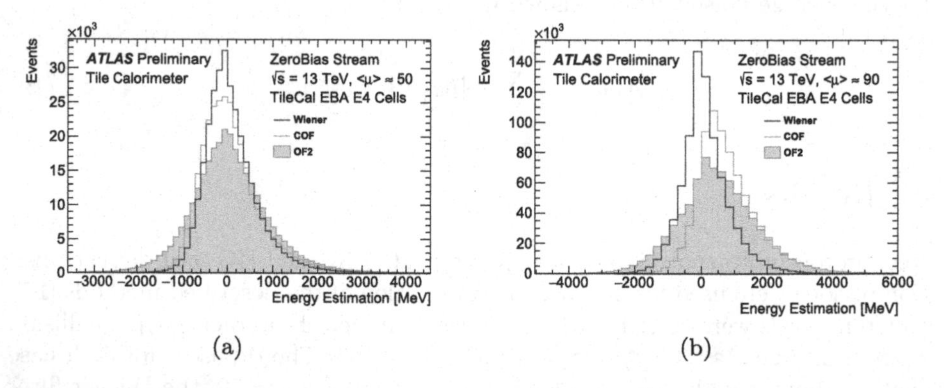

Fig. 3. Reconstructed energy distribution for (a) $<\mu> \approx 50$ and (b) $<\mu> \approx 90$ (extracted from [8]).

The evolution of the mean and RMS of the estimation error distributions according to the $<\mu>$ value are shown in Figs. 4a and b, respectively. Again, the Wiener filter has a lower mean and RMS compared to the COF and OF2 filters in the highest occupancy readout cells. For instance, for $<\mu> \approx 90$, the Wiener filter shows an improvement of approximately 20% and 35% in terms of RMS compared to the COF and OF2 filters, respectively.

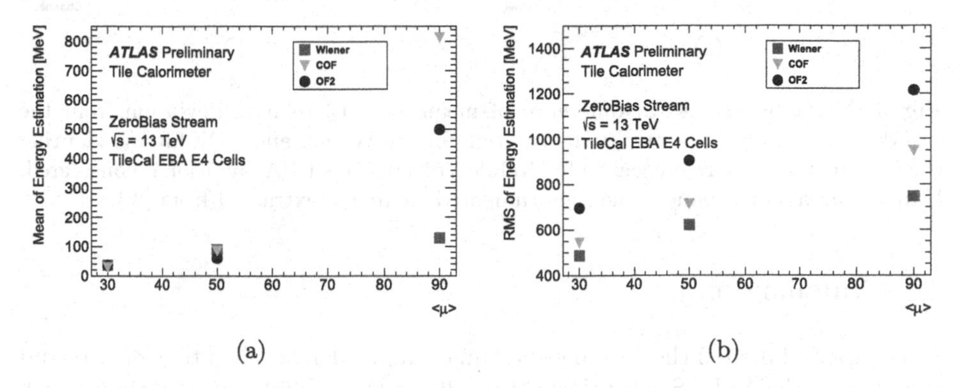

Fig. 4. Evolution of the (a) mean and the (b) standard deviation of reconstructed energy distribution with collision data (extracted from [8]).

In order to verify the estimation performance in other TileCal channels, a complete module was used (Module 01 of the EBA partition). In this analysis, the bias and variance of the three methods are compared. Figure 5a shows the difference in the estimated energy distribution mean between the Wiener filter and COF, using the OF2 method as a reference. Positive values represent higher mean values than OF2 and negative values represent lower values. For example, considering channel 1 (E4 cells), the reference method (OF2) presents a larger mean error with respect to the Wiener Filter (see Fig. 4a), which produces the negative values for the data-points shown in Fig. 5a.

Figure 5b shows the relative percentage difference of the standard deviation of the energy distribution of the methods, adopting the OF2 as a reference and considering only one module. Once again, it is noted that the most significant improvements achieved by using the Wiener filter with respect to COF and OF2 are visible for channels 0 and 1 (cells of highest occupancy in the module). For the other channels, the COF method proved to be more efficient, surpassing the Wiener and OF2 filter, improving the RMS by approximately 25% with respect to the OF2 method.

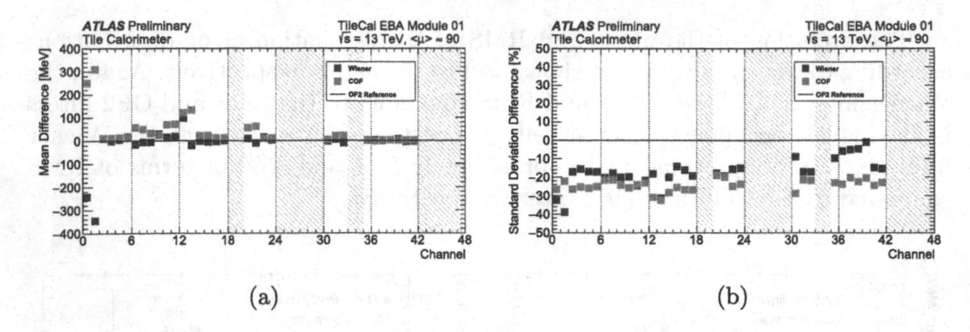

(a) (b)

Fig. 5. Variation of the (a) difference of mean and (b) relative deviation from the standard deviation of the energy distribution for the Wiener and COF filters, adopting the OF2 method as a reference. Only Module 1 of TileCal's EBA partition is considered. Hatched areas correspond to non-instrumented channels (extracted from [8]).

5 Conclusions

This paper addressed the parameter estimation problem applied to energy reconstruction in the ATLAS Tile calorimeter cells under high pile-up conditions which introduce new challenges for the energy estimation task.

Using collision data for severe pile-up conditions ($<\mu> \approx 90$), the Wiener filter method shows an improvement of approximately 35% on the uncertainty of the energy reconstruction compared to the currently used method (OF2), in the most occupied TileCal cells. It was also observed that for low and medium occupation ranges, the COF method presents the best performance for energy estimation, improving the estimation variance by approximately 25% with respect to OF2.

The impact of the different energy reconstruction techniques on the physics objects reconstruction is under study.

Acknowledgments. The authors thanks to CNPq, RENAFAE (MCTIC), FAPERJ, FAPEMIG and FAPESB for the financial support. This study was financed in part by the Coordenação de Aperfeiçoamento de Pessoal de Nível Superior - Brasil (CAPES) - Finance Code 001.

References

1. Adzic, P., et al.: Reconstruction of the signal amplitude of the CMS electromagnetic calorimeter. Eur. Phys. J. C **46S1**, 23–35 (2006). https://doi.org/10.1140/epjcd/s2006-02-002-x
2. Anderson, K., et al.: Design of the front-end analog electronics for the ATLAS tile calorimeter. Nucl. Instrum. Methods Phys. Res., Sect. A **551**(2–3), 469–476 (2005)
3. Filho, L.M.d.A., Peralva, B.S., de Seixas, J.M., Cerqueira, A.S.: Calorimeter response deconvolution for energy estimation in high-luminosity conditions. IEEE Trans. Nucl. Sci. **62**(6), 3265–3273 (2015)

4. ATLAS Collaboration: The ATLAS Experiment at the CERN Large Hadron Collider. JINST **3**, 437 (2008). https://doi.org/10.1088/1748-0221/3/08/S08003
5. ATLAS Collaboration: Readiness of the ATLAS Tile calorimeter for LHC collisions. Eur. Phys. J. C **70**, 1193–1236 (2010)
6. Barbosa, D.P., Filho, L.M.d.A., Peralva, B.S., Cerqueira, A.S., de Seixas, J.M.: Sparse representation for signal reconstruction in calorimeters operating in high luminosity. IEEE Trans. Nucl. Sci. **64**(7), 1942–1949 (2017)
7. CERN: Tile Calorimeter Public Plots, April 2013. https://twiki.cern.ch/twiki/bin/view/AtlasPublic/ApprovedPlotsTile. Accessed 26 May 2021
8. CERN: Tile Calorimeter Public Plots for Collision Data, January 2020. https://twiki.cern.ch/twiki/bin/view/AtlasPublic/TileCaloPublicResults. Accessed 26 May 2021
9. Delmastro, M.: A stand-alone signal reconstruction and calibration algorithm for the ATLAS electromagnetic calorimeter. In: 2003 IEEE Nuclear Science Symposium. Conference Record (IEEE Cat. No. 03CH37515), vol. 2, pp. 1110–1114. IEEE (2003)
10. Evans, L.R., Bryant, P.: LHC machine. JINST **3**, 164 (2008). https://doi.org/10.1088/1748-0221/3/08/S08001
11. Fullana, E., et al.: Digital signal reconstruction in the ATLAS hadronic tile calorimeter. IEEE Trans. Nucl. Sci. **53**(4), 2139–2143 (2006)
12. Haykin, S.O.: Adaptive Filter Theory. Pearson, London (2013)
13. Herr, W., Muratori, B.: Concept of luminosity (2006). https://doi.org/10.5170/CERN-2006-002.361
14. Klimek, P.: Quality factors in TileCal and out-of-time Pile-up, October 2011. http://cds.cern.ch/record/1392389
15. Knoll, G.F.: Radiation Detection and Measurement. Wiley, New York (2010)
16. Marshall, Z.: Simulation of pile-up in the ATLAS experiment. J. Phys. Conf. Ser. **513**(2), 022024 (2014). https://doi.org/10.1088/1742-6596/513/2/022024
17. Martínez, A.R.: The run-2 ATLAS trigger system. J. Phys. Conf. Ser. **762**, 012003 (2016). https://doi.org/10.1088/1742-6596/762/1/012003
18. Oliveira Damazio, D.: Signal processing for the ATLAS liquid argon calorimeter : studies and implementation. Technical report ATL-LARG-PROC-2013-015, CERN, Geneva, November 2013. https://cds.cern.ch/record/1630826
19. Oliveira Goncalves, D.: Energy reconstruction of the ATLAS Tile Calorimeter under high pile-up conditions using the Wiener Filter. Technical report ATL-TILECAL-PROC-2019-002, CERN, Geneva, May 2019. https://cds.cern.ch/record/2674807/
20. Oppenheim, A.V.: Discrete-Time Signal Processing. Pearson Education India, Delhi (1999)
21. Pastore, F.: The ATLAS trigger system: past, present and future. Nucl. Particle Phys. Proc. **273–275**, 1065–1071 (2016). 37th International Conference on High Energy Physics (ICHEP). https://doi.org/10.1016/j.nuclphysbps.2015.09.167
22. Schmidt, B.: The high-luminosity upgrade of the LHC: physics and technology challenges for the accelerator and the experiments. J. Phys. Conf. Seri. **706**, 022002 (2016). https://doi.org/10.1088/1742-6596/706/2/022002
23. Wigmans, R.: Calorimetry: Energy Measurement in Particle Physics. Oxford University Press, Oxford (2017)

Fast Algorithm for Dyslexia Detection

Boris Nerusil, Jaroslav Polec, and Juraj Skunda(✉)

Institute of Multimedia ICT, Slovak University of Technology in Bratislava, Bratislava, Slovakia
juraj.skunda@stuba.sk

Abstract. This article describes a method for detection of cognitive defects based on eye tracking during reading. The aim of this research is to pursue and extend the experiments conducted in Sweden by introducing Conventional signal theory. The dataset used for experiment was acquired by the authors of the research article Screening for Dyslexia Using Eye Tracking during Reading. The provided data consist of 185 subjects divided into two groups. The first group comprises of 88 low risk (LR) subjects and the second group comprises of 97 high risk (HR) subjects. Our measurements achieved a classification accuracy score 86.164% by classifying the subjects into the correct groups.

Keywords: Dyslexia · Eye tracking · KNN · Conventional signal theory

1 Introduction

Using tracking eye movements during reading we are able to create the path of visual attention. Cognitive defects in a subject can be detected by detailed investigation of visual path. Various types of cognitive disorders as autism, schizophrenia, dementia can be identified by using eye tracking. Important scientific publications that focuses on modeling human visual attention have been published since 1988 [1]. Early diagnosis of cognitive disorders is essential in order to detect them at an early stage and begin the treatment. Currently, there are several active researches dealing with visual attention of subjects regarding various cognitive disorders [2–4]. Tracking eye movement of the individual is used in the detection of cognitive disorders, which form basic indicators of cognitive processes. For tracking eye movements and regions of interest of the pictures are commonly employed, for example ROR pictures were used for the identifying the subjects with schizophrenia [2]. In the other cases for the detection of cognitive disorders in subjects with autism or dementia, the text was presented to subjects for reading and based on eye movements the differences between individuals with and without detected dyslexia [6, 7]. This article is specifically dedicated to dyslexia. Eye movements in subjects with dyslexia differ from eye movements in subjects without dyslexia. Individuals with dyslexia require more time to decode particular words and it results in longer and more frequent fixation periods leading to shorter saccades. Fixations are defined as a state when eyes remain steady at least for 50 ms, saccades are determined as movements above threshold distance [4]. We decided to omit the division of eye movements and the input for our classification represents the coordinates of vector of the whole

G. Rozinaj and R. Vargic (Eds.): IWSSIP 2021, CCIS 1527, pp. 152–160, 2022.
https://doi.org/10.1007/978-3-030-96878-6_14

scanned text. Our research is aimed on the experiment, to proof if it is possible to sort out and classify the subjects through Conventional signal theory and if this method is sufficient for the evaluation of acquired data from eye tracker. The article is organized as follows. The first part includes a description of our approach, designed block scheme and displayed vectors view of selected subjects from LR (low risk) and HR (high risk) group and also magnitude spectrum serving for classification of subjects by classifier. In the second part we describe the experiment and the dataset. The final part is devoted to results of experiment, the comparison with already published research, conclusion, acknowledgment and references.

2 Proposed Approach

It was assumed that the data input of coordinates from the right eye (X) and the left eye (Y) which represent the averaged values of coordinates (1) and (2) eye movements of the left eye in the x and y directions (R_x, L_x, R_y, L_y).

$$X = \frac{R_x + L_x}{2} \tag{1}$$

$$Y = \frac{R_y + L_y}{2} \tag{2}$$

A block scheme of the designed system is displayed in Fig. 1. In general the sequences are differently long because subjects had various reading speed, that is why the data were interpolated by DCT3 base function to have the same length,

$$DCT3_{U_{k,n}} = \sqrt{\frac{2}{N}} c_n cos\left(\frac{\pi((2k+1)n)}{2N}\right) \tag{3}$$

$k, n = 0, 1, \ldots, N-1$,

where k represents order in the spectrum and n order in time. Subsequently, the values of the ratio were calculated between original and aligned length. The coordinates in x and y axis were multiplied with the given values of the corresponding subject. The x axis was cut down to particular value, as it is shown in Figs. 2 and 3. Because subjects paid attention to other points after reading the text that eye tracking captured and the redundant information formed approximately half of the total information.

The coordinates of vector are different for subject LR and HR during comparing the amplitudes of the subjects over the time. Fluent course is visible on the x axis of subject 3 LR which is displayed as saw-tooth course and the amplitude is proportional to the line length. Whilst the subject 93 HR does not recognize the course, the eye movement along the x axis is unpredictable during reading particular lines. The eye movement in the horizontal direction was sufficient the most (x axis), that is why vertical eye movement was excluded (y axis). The Discrete Fourier Transform (DFT) was applied to the coordinates of vector of the x axis. The half of the magnitude spectrum of all subjects from LR and HR group was an input into the classifier, Figs. 4 and 5. DC component was omitted from the magnitude spectrum.

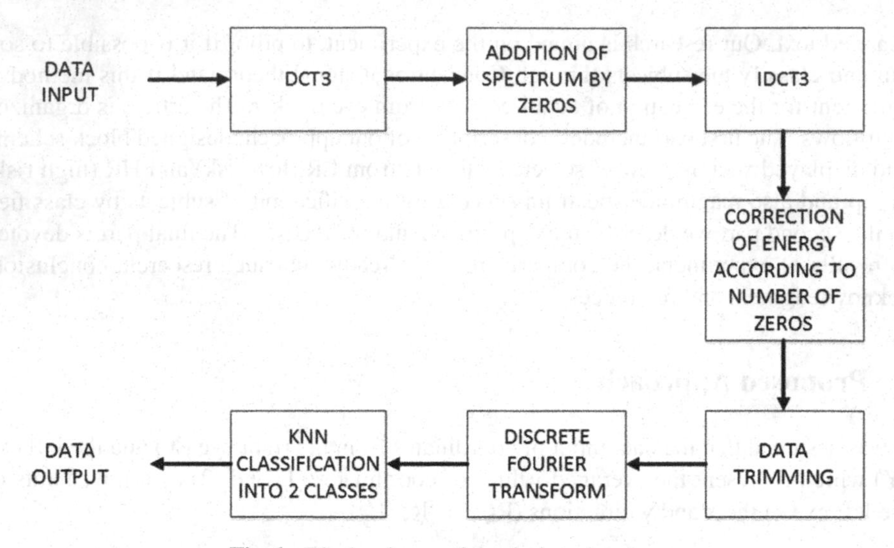

Fig. 1. Block scheme of the designed system.

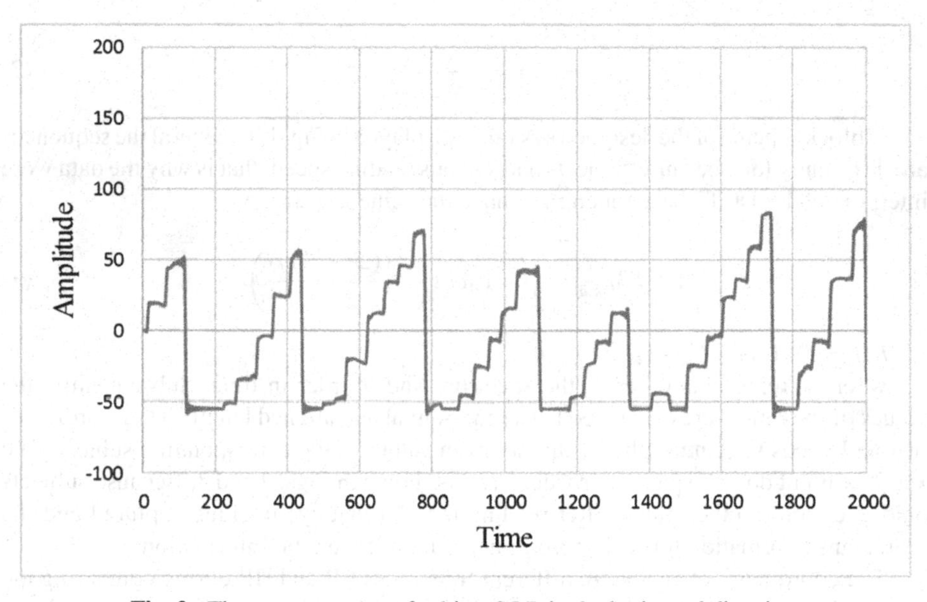

Fig. 2. The eye movement of subject 3 LR in the horizontal direction.

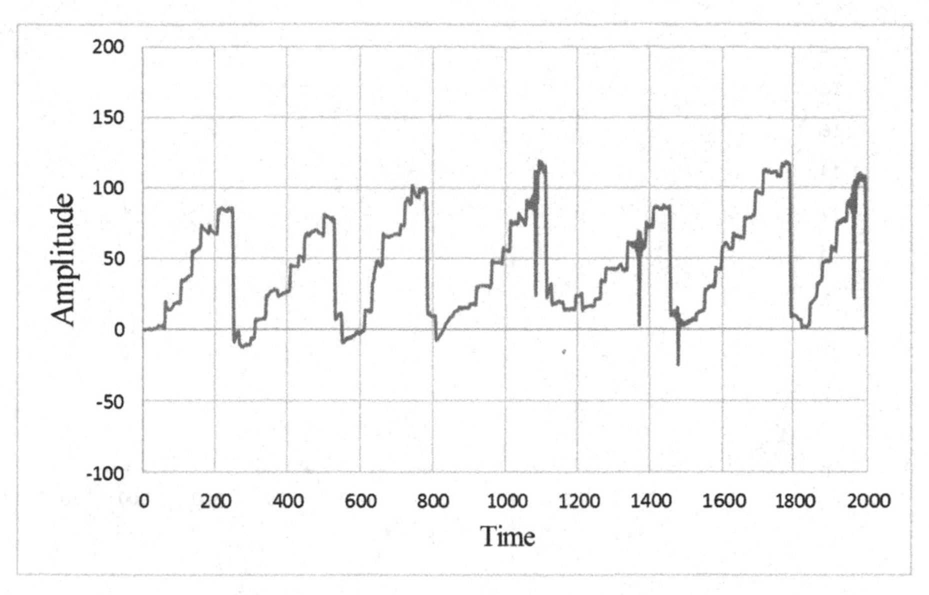

Fig. 3. The eye movement of subject 93 HR in the horizontal direction.

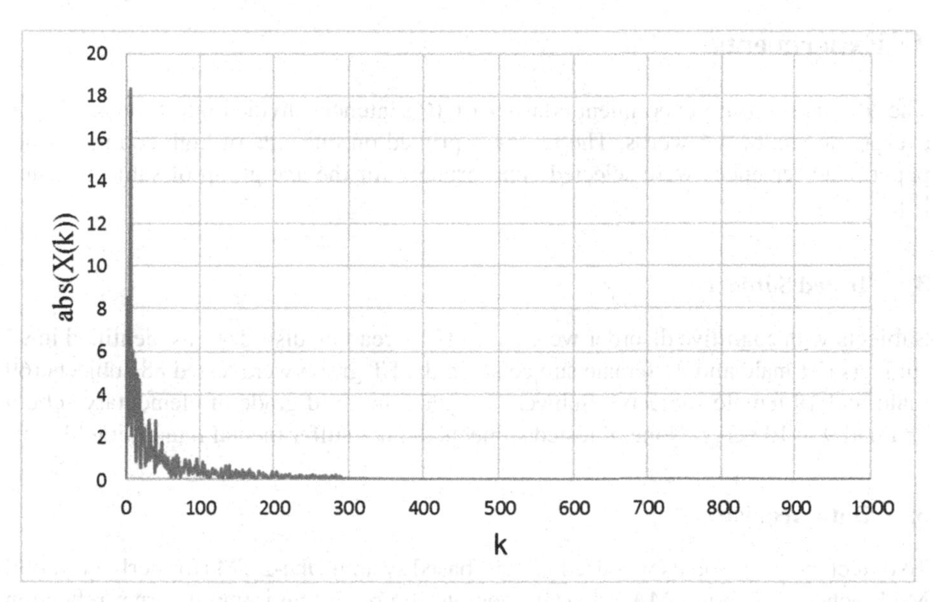

Fig. 4. Magnitude spectrum of subject 3 LR – input into the classifier.

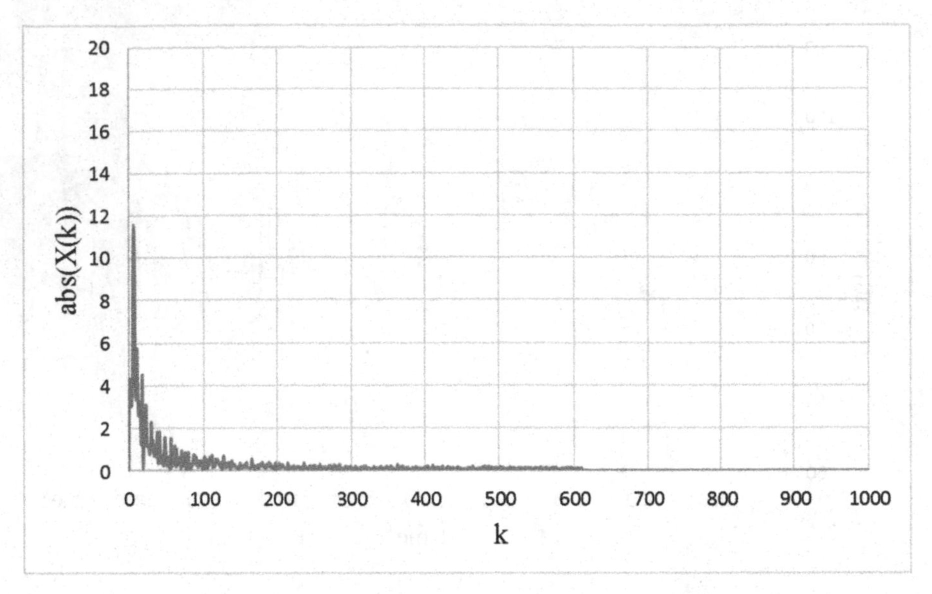

Fig. 5. Magnitude spectrum of subject 93 HR – input into the classifier.

3 Experiments

The text used during experiment consists of 10 sentences divided into 8 lines with an average length of 4.6 words. The text was printed on one side of high contrast white paper. The sentences were selected appropriately for the age group of subjects tested [4].

3.1 Tested Subjects

Subjects with cognitive disorder were in the HR – reading disorder was identified in 97 subjects (71 male and 21 female subjects). In the LT group were tested 88 subjects (69 male and 19 female subjects). Subjects attended the third grade of elementary school and had 9 to 10 years. None of tested subjects did not suffer mental retardation [4].

3.2 Data Acquisition

To detect eye movements was used goggle-based system Obe-2 TM (formerly Permobil Meditech Inc., Woburn MA,). System operates on recording infrared corner reflection of subject. The eye movements were recorded in horizontal and vertical direction at frequency 100 Hz [4]. The eye movements were averaged for our research and only horizontal eye movements were taken into consideration. Authors of published research [4] tracked the changes in the visual path and eye movements of individuals – fixations, saccades, sweeping movements and transients.

3.3 Process of Experiment

The k-nearest neighbors (KNN) algorithm was used for data classification. The tested subject is classified into the appropriate group in this method according to the k-nearest neighbors that influence the subject. The data were divided into two groups of training subjects – 184 subjects and 1 test subject. In the first step, we selected a specific subject from the training data group that served as the test subject. This cycle was repeated 185 times to test each subject for k-nearest neighbors. The correlation coefficient was chosen as metric. After data classification we calculated the positive predictive value PPV – precision (4), TPR – true positive rate, the recall or sensitivity (5), TNR – true negative rate, specificity (6), F1 score (7) and ACC - accuracy (8) for the HR group with dyslexia.

$$PPV = \frac{TP}{TP + FP} \tag{4}$$

$$TPR = \frac{TP}{TP + FN} \tag{5}$$

$$TNR = \frac{TN}{TN + FP} \tag{6}$$

$$F1 = 2 \times \frac{PPH \times TPR}{PPH + TPR} \tag{7}$$

$$ACC = \frac{TP + TN}{TP + TN + FP + FN} \tag{8}$$

4 Results

Several series of calculations with different value of k neighbors was conducted in the KNN classification. K = 5, 19, 25, 33 a 55. The results show us an overview which of the selected number of k-nearest neighbors provides the best classification result. The highest ACC was achieved at K = 33 and K = 25, where accuracy of correct subject classification was 87.03%. The best score for TPR attained 90.72% at K = 33. For TNR the highest reached value was 84.09% at K = 25 and the lowest value was at K = 5, 79.55%. The highest score F1 reached at K = 33 was 88% and the lowest score reached at K = 5 was 86.14%, all results are in Table 1.

Table 1. PPV, TPR, TNR, F1 and ACC score.

	PPV [%]	TPR [%]	TNR [%]	F1 [%]	ACC [%]
K = 5	82.86	89.69	79.55	86.14	84.86
K = 19	85.29	89.69	82.95	87.43	86.49
K = 25	**86.14**	89.69	**84.09**	87.87	**87.03**
K = 33	85.44	**90.72**	82.95	**88**	**87.03**
K = 55	85	87.63	82.95	86.29	85.41

Recall and precision curves use different probability thresholds and through this they summarize for a predictive model the trade-off between the true positive rate and the positive predictive value, Fig. 6.

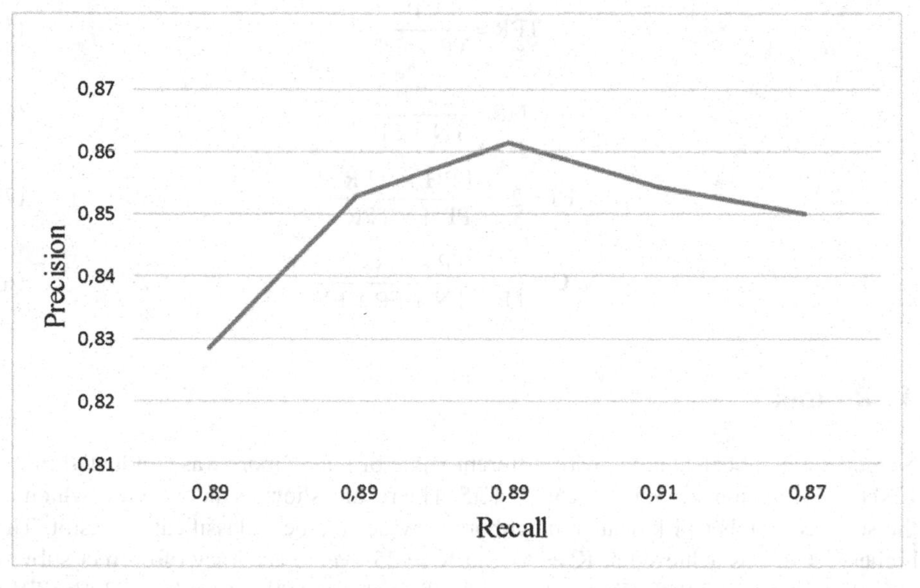

Fig. 6. Recall precision curve for K = 5, 19, 25, 33, 55.

Compared to the results obtained in the previously published research, the authors took into consideration various eye movements and not the whole coordinates of vector of the scanned text, the overall classification accuracy by using SVM-RFE 95.6% ± 4.5. In our case the overall accuracy reached 86.164% with standard deviation ±0.88 (Table 2).

In our study, we focused on the correct classification of the individuals in the HR group, where the highest precision classification achieved 90.72%, it means that 88 out of 97 subjects were classified properly in the KNN, when K = 33 and the lowest precision of classification was achieved when K = 55, 85 subjects out of 97 were classified with accuracy 88%. In cases, where misclassification of different subjects is penalized

Table 2. Comparison of methods.

	KNN	SVM-RFE
Accuracy of detection [%]	86.164 (±0.88)	95.6 (±4.5)

differently, it is not enough to minimize the number of misclassified objects. In medical diagnostics, the misclassification of healthy patient among ill subjects, initiated treatment would have small side effects and it is less dangerous than not providing treatment to seriously ill subject. That means that cost of incorrect classification to the HR group is lower than the cost of incorrect classification to the LR group [5].

5 Conclusion

In comparison to SVMFRE our method propose simpler approach to classification of subjects, due to faster processing of data (only direction of the movement of eyes). The advantage of our system is that the input is only coordinates of vector and there are no other required definitions of events expressing eye movements (number of fixations, length of fixations or length of saccades) and subsequently their selection for vector of features for classification. The answer for our question, whether the Conventional signal theory is suitable for detection reading disorders, is yes, it is. To increase the accuracy it is necessary to analyze the influence and order of shortening the input sequence, its interpolation and the influence of establishing additional relevant feature in the feature vector for classification. Furthermore, it will be interesting to track how another type of classifier will deal with magnitude spectrum of preprocessed signals, for example deep neural network. Whilst this type of reading analysis appears to be appropriate diagnostic methods for other disorders, it might be interesting to monitor if Conventional signal theory would provide good results.

Acknowledgment. I would like to thank the authors for providing the dataset and information regarding the previous research Screening for Dyslexia Using Eye Tracking during Reading. Mattias Nilsson Benfatto, Gustaf Orqvist Seimyr, Jan Ygge, Ten Pansell, Agneta Rydberg, Christer Jacobson.

This research was supported by the VEGA 1/0440/19 research grant.

References

1. Itti, L., Koch, C., Niebur, E.: A model of saliency-based visual attention for rapid scene analysis. IEEE Trans. Pattern Anal. Mach. Intell. **20**(11), 1254–1259 (1998)
2. Polec, J., et al.: Detection of schizophrenia spectrum disorders using saliency maps. In: AICT 2017: 11th IEEE International Conference on Application of Information and Communication Technologies, Moscow, Russia, September 20–22, 2017, pp. 398–402. IEEE, Piscataway (2017)

3. Rello, L., Ballesteros, M.: Detecting readers with dyslexia using machine learning with eye tracking measures. In: W4A '15: Proceedings of the 12th Web for All Conference, International Web for All Conference Florence Italy May, 2015, Article No.: 16, pp. 1–8
4. Benfatto, M.N., Seimyr, G.Ö., Ygge, J., Pansell, T., Rydberg, A., Jacobson, C.: Screening for dyslexia using eye tracking during reading. PLoS ONE **11**(12), e0165508 (2016)
5. Šikudová, E., Černeková, Z., Benešová, W., Haladová, Z., Kučerová, J.: Computer Vision Object Detection and Recognition. Wikina, Praha (2013)
6. Fraser, K.C., Fors, K.L., Kokkinakis, D., Nordlund, A.: An analysis of eye-movements during reading for detection of mild cognitive impairment. In: Proceedings of the 2017 Conference on Empirical Methods in Natural Language Processing, pp. 1016–1026
7. Yaneva, V., Ha, L.A., Eraslan, S., Yesilada, Y., Mitkova, R.: Detecting autism based on eye-tracking data from web searching tasks. In: W4A '18: Proceedings of the Internet of Accessible Things, The Internet of Accessible Things Lyon France, April 2018, Article No.: 16 , pp. 1–10

Automatic Recognition of Native Advertisements for the Slovak Language

Vanesa Andicsova, Zuzana Bukovcikova(✉), Dominik Sopiak, and Milos Oravec

Faculty of Electrical Engineering and Information Technology, Slovak University of Technology in Bratislava, Ilkovicova 3, 812 19 Bratislava, Slovak Republic
{xandicsova,xbukovcikova,dominik.sopiak,milos.oravec}@stuba.sk

Abstract. In recent years the native advertisement is becoming more and more prevalent in online spaces. Differentiating between genuine content and native advertisement using Natural Language Processing is therefore also becoming a very interesting research topic. In this paper, we examine the possibilities of using deep textual representation for the Slovak language to recognize the "PR (Public relations) articles" (that serve as a native advertisement in this context) from authentic news articles on popular Slovak news websites. We show that the BERT (Bidirectional Encoder Representations from Transformers) embeddings as a text representation are sufficient for this task (achieving accuracy over 80% even with a statistical model - Logistic Regression) and that the models generally perform better without prior lemmatization.

We have scraped three Slovak news websites (for a total of 5455 news articles containing both paid-for content and a wide variety of genuine categories), and we have evaluated multiple binary classification methods (Logistic Regression, Random forest classifier and Support Vector Machines) trained on top of generated RoBERTa sentence embeddings. On our testing set, we were able to achieve an accuracy of 85.13%.

Keywords: NLP · Slovak language · Native advertisement · Text classification

1 Introduction

The topic of document classification is a very popular field of Natural Language Processing (NLP), mainly focusing on the categorization based on topic, sentiment, or entailment [11]. However, one could argue that the detection of genuine (or authentic) content from paid content is a topic that also deserves the attention of researchers, mainly since the rise of native advertisement in recent years. The term native advertisement (later also referenced as "PR content" or "paid-for content") was first introduced in 2011, and ever since then, it has infiltrated almost every big internet platform, including digital news websites.

This paper presents a method on how to distinguish between native advertisements and genuine news articles - using BERT multilingual model [16] for text

G. Rozinaj and R. Vargic (Eds.): IWSSIP 2021, CCIS 1527, pp. 161–171, 2022.
https://doi.org/10.1007/978-3-030-96878-6_15

encoding and then training a classifier to distinguish between real and paid-for content.

The rest of the paper is organized as follows: in the next subsections, we briefly introduce the concept of native advertisement, the motivation behind the paper, and related works. In Sect. 2 we discuss the properties of our digital news dataset used for training and testing. In Sect. 3 the methodology of our experiments is summarized. Then, in Sects. 4 and 5 we present and analyse results achieved on our testing set.

1.1 Native Advertisement

Native advertisement in digital newspapers is a type of advertisement trying to be indistinguishable from the actual content being published - in formatting, tone, and to an extent, even content.

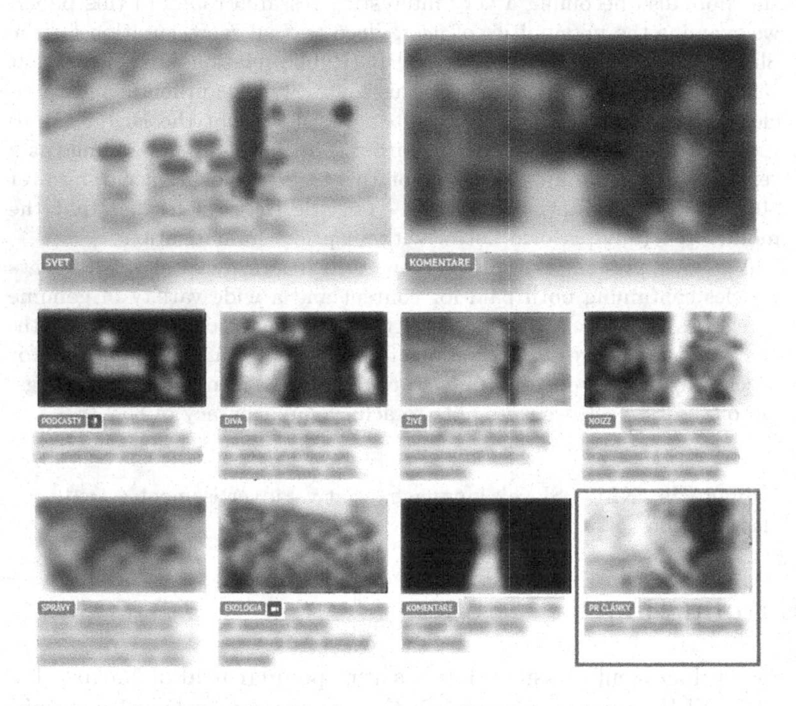

Fig. 1. A native advertisement placed among authentic articles on the news page website.

Sponsored news articles have also been on the rise in recent years and are often listed among news website's genuine content. They aim to provide exciting or entertaining information to the reader and "connect" with him, sometimes including brand name or hyperlink only in the later parts of the article. Even though disclosure that the article belongs among a sponsored content is usually present on the news websites, its position on the website can confuse human readers and possibly skew results of automated media analysis (example of PR articles placement is shown in Fig. 1).

The topic of recognizing native advertisements is also relevant because of the amount of native PR content appearing on the Slovak websites. According to IAB Slovakia - an association for online advertising in Slovakia - in 2020, the total cost of native advertising makes up 4% of all advertisement costs in Slovak online space[1], and the popularity is steadily rising (14.26% increase in cost between 2019 and 2020).

1.2 Related Work

The problem of document classification using deep representation has received much attention ever since deep learning became popular for natural language processing tasks. It has quickly replaced statistical methods as the state-of-the-art for many text classification tasks.

With deep learning, it seems that over the years the focus has shifted from generating representation for words (e.g. Word2Vec [10] or GloVE [13]) to generating embeddings for whole sentences or texts. Many architectures were explored for this task; however, prevalent (and accurate for many NLP tasks) are models using Transformers architecture [20], which consists of series of attention-based blocks. This architecture allows for self-supervised training and variable-length input.

BERT (Bidirectional Encoder Representations from Transformers) is a model for sentence encoding based on Transformers architecture introduced in 2018 [4], which can be considered state-of-the-art embedding model [11]. BERT models generally consist of stacks of Encoder and Decoder layers, composed of a self-attention layer, feed-forward layer, and a residual skip connection.

There have been multiple variants of BERT introduced after 2018, for example ALBERT [8] (a lightweight version of BERT with smaller memory consumption and faster training), or DistilBERT [17] (also smaller and faster, but pre-trained through knowledge distillation).

An interesting extension is the SBERT (Sentence BERT) [15], where a BERT model is fine-tuned in a siamese/triplet architecture - this model is computationally efficient and also produces embeddings that can be compared using cosine similarity reflecting semantical meaning. Important extension has also been RoBERTa [9], since it is a multilingual model capable to perform well on low-resource languages.

While there are many papers published and models available for texts in English, the topic of NLP in other languages is limited, however gaining traction in recent years [16,22]. In this paper we are focusing on the Slovak language, for which deep representation was previously used for sentiment classification [12], document summarization [18] or punctuation restoration [5]. In [18] the

[1] Report available at: https://www.iabslovakia.sk/vydavky-do-reklamy/vydavky-do-internetovej-reklamy-2020-sk/ (Last accessed 09 Apr 2021).

researchers also present a Slovak summarization dataset build from scraping Slovak news websites.

The topic of native advertisement is also attracting researchers' attention since its surge in 2011 [19]. The problem of detecting the paid-for content was studied in [14], where the authors build a system to detect native advertisements and other extraneous content in podcast episodes, also using BERT multilingual model on episodes transcript.

2 Dataset

Since NLP datasets for the Slovak language are limited (especially with the connection to advertisements), we have built our own. The dataset that was used for experiments in this paper was collected using web scraping from three Slovak news websites[2]. We have tried to build the collection by choosing the subcategory of authentic articles that the native advertisements most commonly try to emulate - for example, celebrity news (since in the paid-for articles, it is often a celebrity promoting a product) or health news (self-help angle is also often present).

Altogether there were 3000 genuine and 2455 PR articles, which were later used for training our models. The details of our scraped database are summarized in Table 1. The publishing date of articles from websites ranges from November 2006 to March 2021.

Since the information about whether an article is PR content is disclosed on the scraped websites (and they can be found in their own subsection), we were able to assign labels for the documents automatically. Some mislabelling was present on the news websites, and it was manually corrected to the best of our ability.

3 Methodology

After acquiring the dataset, we have built a system for training our model for binary classification. This process can be split into three steps:

- **preprocessing of texts** consisting of text cleaning, splitting into parts and lemmatization,
- **deep representation** using BERT multilingual model to generate embeddings for different parts of scraped articles,
- **binary classification** used on top of our generated embeddings, using both statistical and machine learning methods.

A more detailed description of these steps is in the following subsections.

[2] Available at URLs: https://www.aktuality.sk/, https://www.cas.sk/, https://techbox.dennikn.sk/ (Last accessed 09 Apr 2021).

Table 1. Properties of scraped dataset.

Source	Total number of articles	Section	Authentic content	PR content
www.aktuality.sk	1759	Home news Economics Health Culture PR content	1000	759
www.cas.sk	2499	PR content Culture and celebrity news Health	1000	1499
www.techbox.sk	1197	Technews PR content	1000	197
Total	**5455**	—	**3000**	**2455**

3.1 Data Preparation

In the first step, three parts from every article on the news websites were extracted:

1. title,
2. introduction (structurally highlighted on the website - usually a first paragraph or a short summary),
3. main content.

We have cleaned all of them from hyperlinks, scripts, and all the formatting to get only the pure text from all the relevant tags. We have then checked the language of scraped articles.

Then, we have lemmatized the texts using open-source model[3] from SparkNLP library [6]. Using these steps, we have created eight different possible texts for every document, which were later used for our experiments and are summarized in Sect. 4.

For better understanding of the dataset and our problem, we include Table 2, where we show some exemplary titles from PR articles.

3.2 Deep Representation

Next we have generated the embeddings for the text using BERT multilingual model[4] [16].

[3] Available at https://nlp.johnsnowlabs.com/2020/05/05/lemma_sk.html (Last accessed 09 Apr 2021).

[4] Known as *stsb-xlm-r-multilingual*; available at: https://huggingface.co/sentence-transformers/stsb-xlm-r-multilingual (Last accessed 09 Apr 2021).

Table 2. Sample titles of paid-for articles in our database (english translation provided by authors).

Sample titles of PR articles
[*Original*] Chcete tankovať lacnejšie? Máme pre Vás skvelý tip! [*Lemmatized*] chcete, tankovať, lacný, ?, máme, pre, vás, skvelý, tip, ! [*English translation*] Do you want to pay less for gas? We have a tip for you!
Original] Zabudnite na sv. Valentína: Toto sú oslavy lásky, o ktorých ste ani netušili! [*Lemmatized*] zabudnite, na, sv,., valentín, :, toto, byť, oslava, láska, ,, o, ktorý, byť, ani, netušiť, ! [*English translation*] Forget about Valentine's day: Here are love celebrations which you have never heard about before!
[*Original*] Čo všetko musí absolvovať práčka, kým sa dostane až ku vám domov? [*Lemmatized*] čo, všetko, musieť, absolvovať, práčka, ,, kto, sa, dostať, až, k, vy, dom, ? [*English translation*] What does a washing machine have to go through before it gets to your home?

This model has been trained using a teacher-student architecture, where a teacher model is monolingual (in this case using SBERT architecture [15]) and a student model is multilingual (using XLM-RoBERTa [3]). The training process tries to map all the embeddings from the student model (for multiple languages) to be similar to the embedding of a teacher model. This model creates a 768 dimensional representation for every input text (of variable length).

Using this architecture, the student model (with 12 layers and 12 heads) creates a representation for Slovak texts, which should have the same advantages as the original SBERT embeddings (mainly semantic cosine-similarity). The models were trained on corpora from SNLI [1], MultiNLI [21] and STS benchmark train set [2].

Considering our experiments with prior lemmatization, an important note for the student embedding model is that it starts with SentencePiece tokenization - an unsupervised data-driven and language-independent tokenizer[5] [7]. Using this tokenization, a sentence "*Slovensko zasiahla vlna obrovského nárastu dopytu po produktoch na kĺby!*" (an exemplary title of a PR article from dataset) would be split into tokens "[*'_Slovensko', '_za', 'sia', 'hla', '_v', 'lna', '_obrovské', 'ho',*

[5] Available at https://github.com/google/sentencepiece (Last accessed 09 Apr 2021).

'_ná', 'rast', 'u', '_dop', 'ytu', '_po', '_produkto', 'ch', '_na', '_k', 'ĺ', 'by', '!']", which will then be processed by the RoBERTa model. This tokenization keeps punctuation and also codes inflectional morphology.

Embeddings were computed for every one of the eight possible texts connected to a single article separately.

3.3 Training

After we have generated the embeddings - vector of size 768 - for every text configuration, multiple models for each configuration from preprocessing were trained:

- Logistic Regression,
- Random Forest Classifier,
- and Support Vector Machine Classifier.

We have split the dataset into training and testing subsets using a ratio of 75:25, totaling 4072 samples for training and 1358 for testing. The models were trained using cross-validation with a 5-fold split.

A grid search to find optimal training parameters was also employed for every classifier - type of penalty for logistic regression, cost parameters and kernel type (linear or RBF) for SVMs, and the number of trees and split stopping parameters for random forest classifier.

4 Results

Altogether we have trained 32 models with different input data and machine learning method. Their respective details and accuracies are summarized in Table 3.

As we can see, the best results were achieved using embeddings generated from only the title of the articles coupled with their introduction without prior lemmatization. The best performing model was a Support Vector Classifier with RBF kernel and cost parameter set to 1, while the lowest accuracy was achieved using Random Forest Classifier.

Detailed results (with achieved accuracies also for subcategories of authentic content) for our best performing model are summarized in Table 4. While the resulting dataset size of the respective subcategories in the testing set is not large, these results suggest there is no specific issue with any authentic article subtype and that the model was able to differentiate between PR and authentic content even in the subcategories which may seem more challenging (for example category health).

A couple of interesting notes can be observed from the presented results:

- **Models, in general, perform better without prior lemmatization.** This was previously observed by researchers using BERT models for the Slovak language [12,18]. While lemmatization is still often used in automatic language processing for the Slovak language (especially coupled with statistical

Table 3. Achieved results.

Method	Lemma-tized	Part of text	Category PR		Category authentic		Accuracy
			Precision	Recall	Precision	Recall	
Logistic regression	Yes	Title	0.7470	0.7368	0.8028	0.8111	0.7791
		Introduction	0.7223	0.7470	0.8035	0.7827	0.7673
		Title + Introduction	0.7759	0.7692	0.8265	0.8318	0.8049
		Full text	0.7826	0.7692	0.8276	0.8383	0.8085
	No	Title	0.7530	0.7402	0.8059	0.8163	0.7835
		Introduction	0.7475	0.7538	0.8125	0.8072	0.7842
		Title + Introduction	0.7718	0.7863	0.8360	0.8241	0.8078
		Full text	0.7838	0.7932	0.8420	0.8344	0.8166
Random Forest	Yes	Title	0.7656	0.6308	0.7534	0.8538	0.7577
		Introduction	0.7608	0.6308	0.7526	0.8499	0.7555
		Title + Introduction	0.8251	0.6855	0.7890	0.8900	0.8019
		Full text	0.7888	0.6957	0.7889	0.8590	0.7887
	No	Title	0.7634	0.6564	0.7649	0.8461	0.7644
		Introduction	0.7697	0.6342	0.7557	0.8564	0.7607
		Title + Introduction	0.7895	0.6923	0.7870	0.8603	0.7879
		Full text	0.7940	0.7179	0.8010	0.8590	0.7982
SVM	Yes	Title	0.7772	0.7692	0.8269	0.8331	0.8056
		Introduction	0.7961	0.7675	0.8287	0.8512	0.8152
		Title + Introduction	0.8153	0.800	0.8508	0.8629	0.8358
		Full text	0.8202	0.8034	0.8535	0.8668	0.8395
	No	Title	0.7843	0.7709	0.8289	0.8396	0.8100
		Introduction	0.7983	0.7915	0.8430	0.8473	0.8233
		Title+ Introduction	**0.8296**	**0.8239**	**0.8674**	**0.8719**	**0.8513**
		Full text	0.8280	0.7983	0.8514	0.8745	0.8417

Table 4. Detailed results for our best model.

Subcategory	Size of testing subset	Accuracy
PR article	858	0.8239
Culture and celebrity news	192	0.8802
Home news	67	0.8955
Technews	261	0.8774
Economy	66	0.8636
Health	187	0.8503

analysis of texts), this behaviour is perhaps not surprising (since the authors of the RoBERTa model we have used do not mention using lemmatization while training). Also, it can be theorized that the model performs better with inflectional morphology since it can help assess the tone or subject of the sentence for the Slovak language (and that SentencePiece representation of tokens could have the capacity to capture these inflections).

- **Longer texts do not necessarily yield higher accuracies.** There was only a small observable benefit to encoding full text instead of just title and introduction (first paragraph). For clarification purposes, the minimum, maximum, and mean length of the used texts parts are summarized in Table 5.

Table 5. Length (number of words) of texts used for training.

	Min length	Max length	Mean length
Title	1	25	9.581 ± 3.156
Introduction	4	112	30.922 ± 15.673
Content	7	3791	367.614 ± 275.247

- **Relatively good results can be achieved using only the title of the article.** This was surprising since the titles are usually short (in our dataset, the maximum length was 21 words, while the mean length was only 9.577 words). It can perhaps be explained from the nature of data itself - while the goal of a native advertisement is to appear as close to genuine content as possible, human readers are often capable of determining whether an article is paid-for from the title itself and it would appear that the embeddings carry this sentiment as well. Adding longer texts (providing more context) improves the results further.

5 Conclusion and Future Work

The results in this paper show that the BERT multilingual model performs well for the task of recognition of paid-for content in the Slovak language. We have also observed that the models trained on texts without lemmatization performed better (confirming findings from [12] and [18]), which may not be surprising since the BERT models use SentencePiece tokenization. Also we have found that this problem can be solved using only short texts of the articles (title and/or introduction).

However what may be surprising are the (relatively) high accuracies of all the trained models, since it is the goal of native advertisements to appear as close as possible to the genuine content published on the websites. However, there are some clues by which even human readers can distinguish between true and PR content without explicit labeling (positive tone, direct addressing of the readers, etc.) and it can be hypothesized that the BERT model encodes those well.

For future research, it may be interesting to infer which content is paid for not only from the text but also from different articles' properties. In our experiments, we have striped the text from the hyperlinks or formatting. However, one can hypothesize that those often help human readers distinguishing the content types.

Also, perhaps more informative results can be achieved by further specifying the subtype of PR articles used for training and evaluation since there is a varying level of emulating the genuine content between them.

We believe it is important to research this topic further since the native advertising strategies are becoming more and more popular (they seems to also gain traction on social media)- and often not only for promoting products. It may also become crucial because not all online spaces are disclosing the paid-for content.

Acknowledgment. The research described in the paper was done within the International Center of Excellence for Research on Intelligent and Secure Information and Communication Technologies and Systems - II. stage, ITMS code: 313021W404, co-financed by the European Regional Development Fund.

References

1. Bowman, S.R., Angeli, G., Potts, C., Manning, C.D.: A large annotated corpus for learning natural language inference. arXiv preprint arXiv:1508.05326 (2015)
2. Cer, D., Diab, M., Agirre, E., Lopez-Gazpio, I., Specia, L.: SemEval-2017 task 1: semantic textual similarity multilingual and crosslingual focused evaluation. In: Proceedings of the 11th International Workshop on Semantic Evaluation (SemEval-2017), pp. 1–14. Association for Computational Linguistics, Vancouver, Canada, August 2017. https://doi.org/10.18653/v1/S17-2001, https://www.aclweb.org/anthology/S17-2001
3. Conneau, A., et al.: Unsupervised cross-lingual representation learning at scale (2020)
4. Devlin, J., Chang, M., Lee, K., Toutanova, K.: BERT: pre-training of deep bidirectional transformers for language understanding. CoRR abs/1810.04805 (2018). http://arxiv.org/abs/1810.04805
5. Hládek, D., Staš, J., Ondáš, S.: Comparison of recurrent neural networks for Slovak punctuation restoration. In: 2019 10th IEEE International Conference on Cognitive Infocommunications (CogInfoCom), pp. 95–100 (2019). https://doi.org/10.1109/CogInfoCom47531.2019.9089903
6. Kocaman, V., Talby, D.: Spark NLP: natural language understanding at scale. Softw. Impacts, 100058 (2021). https://doi.org/10.1016/j.simpa.2021.100058
7. Kudo, T., Richardson, J.: SentencePiece: a simple and language independent subword tokenizer and detokenizer for neural text processing. In: Proceedings of the 2018 Conference on Empirical Methods in Natural Language Processing: System Demonstrations, pp. 66–71. Association for Computational Linguistics, Brussels, Belgium, November 2018. https://doi.org/10.18653/v1/D18-2012, https://www.aclweb.org/anthology/D18-2012
8. Lan, Z., Chen, M., Goodman, S., Gimpel, K., Sharma, P., Soricut, R.: Albert: a lite BERT for self-supervised learning of language representations (2020)
9. Liu, Y., et al.: Roberta: a robustly optimized BERT pretraining approach (2019)
10. Mikolov, T., Chen, K., Corrado, G., Dean, J.: Efficient estimation of word representations in vector space (2013)
11. Minaee, S., Kalchbrenner, N., Cambria, E., Nikzad, N., Chenaghlu, M., Gao, J.: Deep learning based text classification: A comprehensive review. arXiv preprint arXiv:2004.03705 (2020)
12. Pecar, S., Simko, M., Bielikova, M.: Sentiment analysis of customer reviews: impact of text pre-processing. In: 2018 World Symposium on Digital Intelligence for Systems and Machines (DISA), pp. 251–256 (2018). https://doi.org/10.1109/DISA.2018.8490619

13. Pennington, J., Socher, R., Manning, C.: GloVe: global vectors for word representation. In: Proceedings of the 2014 Conference on Empirical Methods in Natural Language Processing (EMNLP), pp. 1532–1543. Association for Computational Linguistics, Doha, Qatar, October 2014. https://doi.org/10.3115/v1/D14-1162, https://www.aclweb.org/anthology/D14-1162
14. Reddy, S., Yu, Y., Pappu, A., Sivaraman, A., Rezapour, R., Jones, R.: Detecting extraneous content in podcasts (2021)
15. Reimers, N., Gurevych, I.: Sentence-BERT: sentence embeddings using Siamese BERT-networks. arXiv preprint arxiv:1908.10084 (2019)
16. Reimers, N., Gurevych, I.: Making monolingual sentence embeddings multilingual using knowledge distillation. In: Proceedings of the 2020 Conference on Empirical Methods in Natural Language Processing. Association for Computational Linguistics, November 2020. https://arxiv.org/abs/2004.09813
17. Sanh, V., Debut, L., Chaumond, J., Wolf, T.: Distilbert, a distilled version of BERT: smaller, faster, cheaper and lighter (2020)
18. Suppa, M., Adamec, J.: A summarization dataset of Slovak news articles. In: Proceedings of the 12th Language Resources and Evaluation Conference, pp. 6725–6730 (2020)
19. Taylor, C.R.: Native advertising: the black sheep of the marketing family (2017)
20. Vaswani, A., et al..: Attention is all you need (2017)
21. Williams, A., Nangia, N., Bowman, S.: A broad-coverage challenge corpus for sentence understanding through inference. In: Proceedings of the 2018 Conference of the North American Chapter of the Association for Computational Linguistics: Human Language Technologies, Volume 1 (Long Papers), pp. 1112–1122. Association for Computational Linguistics (2018). http://aclweb.org/anthology/N18-1101
22. Wu, S., Dredze, M.: Beto, Bentz, Becas: the surprising cross-lingual effectiveness of BERT (2019)

Document Filter for Writer Identification

Fabio Pignelli[1(✉)], Luiz S. Oliveira[2], Alceu S. Britto Jr.[3], Yandre M. G. Costa[1], and Diego Bertolini[1,4]

[1] State University of Maringá, Maringá, PR, Brazil
[2] Federal University of Paraná, Curitiba, PR, Brazil
[3] Pontifical Catholic University of Paraná, Curitiba, PR, Brazil
[4] Federal Technological University of Paraná, Campo Mourão, PR, Brazil

Abstract. The writing can be used as an important biometric modality which allows to unequivocally identify an individual. It happens because the writing of two different persons present differences that can be explored both in terms of graphometric properties or even by addressing the manuscript as a digital image, taking into account the use of image processing techniques that can properly capture different visual attributes of the image (e.g. texture). In this work, we perform a detailed study in which we dissect whether or not the use of a dataset with only a single sample taken from some writers may skew the results obtained in the experimental protocol. In this sense, we propose here what we call "Document Filter". The Document Filter protocol is supposed to be used as a preprocessing technique, in such a way that all the data taken from fragments of the same document must be placed either into the training or into the test set. The rationale behind it, is that the classifier must capture the features from the writer itself, and not features regarding other particularities which could affect the writing in a specific document (e.g. emotional state of the writer, pen used, paper type, and etc.). By analyzing the literature, one can find several works dealing with the writer identification problem. However, the performance of the writer identification systems must be evaluated also taking into account the occurrence of writer volunteers who contributed with a single sample during the creation of the manuscript databases. To address the open issue investigated here, a comprehensive set of experiments was performed on the IAM, CVL and BFL databases. They have shown that, in the most extreme case, the recognition rate obtained using the DF protocol drops 30.94% points.

Keywords: Writer identification · Single-sample writer · Document filter · Texture

1 Introduction

Writer identification is an ordinary task, necessary in some specific domains, such as forensic science. The main goal in this task is to define who is the person that wrote a document by handwriting text. Automatic writer identification is a hot topic, intensively addressed by the pattern recognition research community.

G. Rozinaj and R. Vargic (Eds.): IWSSIP 2021, CCIS 1527, pp. 172–184, 2022.
https://doi.org/10.1007/978-3-030-96878-6_16

This task, as the writer identification task, may be considered a quite challenging task, because each writer corresponds to one class in these problems, thus characterizing a problem with numerous classes. In addition, sometimes the writing style of different people presents a reasonable similarity, which makes the inter-class likeness increase proportionally to the number of writers [18].

Another adversity happens because as well as two distinct people do not have identical writing, one person does not reproduce its own writing twice identically [13]. This detail implies in an additional difficulty to identify whether or not two given manuscripts were written by the same person.

By analyzing the literature, we can easily find several experiments attacking the task investigated here with very high performance rates [3,9,22]. However, as we know, in real world problems the scenery found is frequently different from that enforced by many databases used in some typical experimental setup. In many cases, the experimentation is performed using a scarce amount of manuscript samples per writer, what makes difficult to build a more robust classifier model.

With that in mind, in this work we try to open up this "black-box" developing a series of experiments to evaluate a quite subtle issue surrounding the writer identification task. For this purpose, we evaluate the impacts of the introduction of a mandatory restriction which prohibits the use of data taken from different fragments of the same manuscript both on the train and test sets. We call this restriction "Document Filter" or "DF", and it was inspired by a similar protocol introduced by Pampalk et al. [17] in the music genre classification task. In that work, the authors intended to avoid the creation of classifiers able to classify artist ("Artist Filter"), instead of music genre. It was reported that in the most extreme case, the recognition rate reduced from 71% to 27% when that filter was applied.

There are several factors that can help to understand why using a single document from a writer can somehow introduce a bias to the results. It happens because there are some factors that may introduce slight differences in two different documents written by the same person, like the emotional state of the writer; the fluency of the pen/pencil used for writing; and the paper writability (smooth/rough), among others [13].

To demonstrate this hypothesis, we evaluate the difference between performances with or without the DF protocol. Hence, we have two different protocols: i) With samples taken from the same document (i.e. different regions of the same document image) both in train and test sets, and ii) with samples from different documents in the train and test sets. A set of experiments was carried out, and they confirm that, in the most extreme cases, the identification rate obtained using DF can drop from 62.85% to 31.91% on the IAM dataset, from 79.30% to 63.15% on the CVL dataset, and from 67.65% to 46.92% on the BFL dataset.

This paper is organized as follows: in Sect. 2 we present details about the experimental setup; Sect. 3 shows the obtained results; a critical review is described in Sect. 4; finally, we present the concluding remarks in Sect. 5.

2 Experimental Setup

In this section, we describe the databases used to perform the experiments in Subsect. 2.1. Following, we describe in details how we organized the data aiming to get the desired evaluation in Subsect. 2.2. Lastly, we present some information about the feature extraction method used in our experiments Sect. 2.3.

Figure 1 illustrates the proposed method. Figure 1(a) shows the proposed scheme without use of DF and Fig. 1(b) illustrates the use of DF protocol.

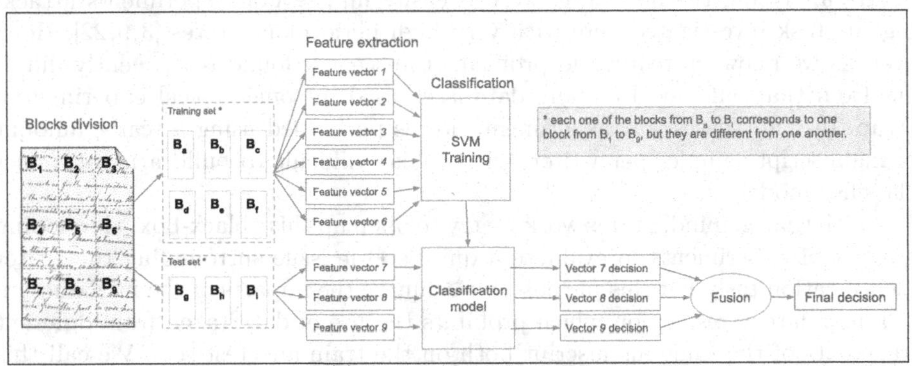

(a) Approach without DF.

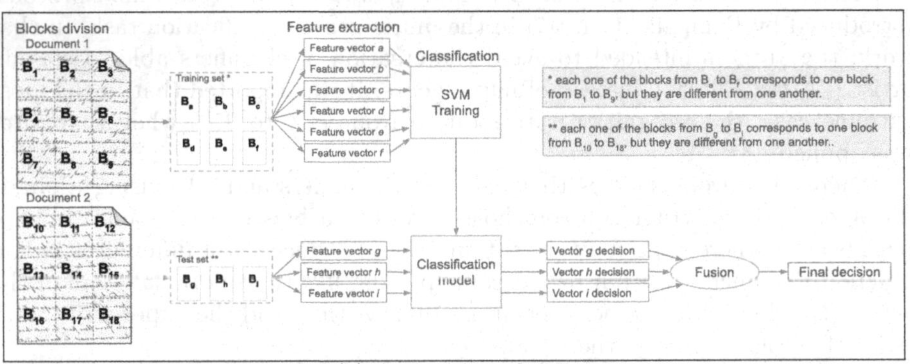

(b) Approach with DF.

Fig. 1. General overview of the proposed/evaluated approach.

2.1 Databases

The databases chosen to perform the experiments are IAM, CVL, and BFL. The IAM database is composed of documents written in English and presents a significant variation regarding the amount of content per manuscript, as it has been created on the text-independent mode, i.e. the volunteers are free to create the manuscript content. The current version of the IAM, presented by Marti and Bunke [14] has 115,320 samples of handwriting words, distributed in 13,353 lines of text, and the 657 writers produced the handwritten content using a lexicon with 10,841 different words. The number of handwritten samples collected per writer varies widely (i.e. from 1 to 59 samples per writer), and the vast majority (356 writers) contributed with only a single sample.

CVL database [12] is a text-dependent database, with contributions of 310 writers. The texts are excerpts from literary works, and they contain between 47 and 90 words. This database can be also employed in multi-script tasks because it contains documents written both in English and German. All writers contributed with at least four documents in English and one in German, totaling 1,604 handwriting samples. The documents were digitalized generating images in RGB color space, with 300 dpi of resolution.

The BFL database, proposed by Freitas et al. [7] is composed of manuscripts taken from 315 writers and contains three document samples per writer collected according to the text-dependent mode (i.e. the writers are asked to copy predefined texts). The texts were written in Portuguese and were carefully chosen, in such a way that they contain all letters, numbers, and special characters from the Portuguese Language. The volunteers who contributed to the creation of the database used their own pen, and the text was written on white paper with no pen-draw baseline. Last, the documents were scanned in gray levels with a resolution of 300 dpi.

The databases chosen to be used in this work, were selected because they have suitable characteristics to support the investigations intended here. Moreover, they have been widely used in other works described in the literature. It is

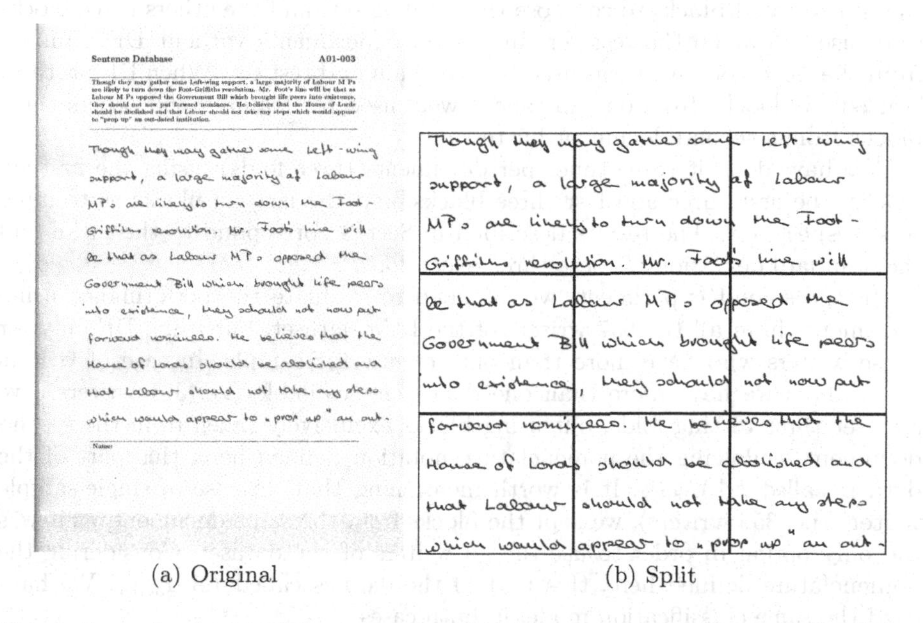

(a) Original (b) Split

Fig. 2. Example of original document from IAM database and split preprocessed document in nine blocks.

important to observe that we had to discard some samples for many writers, as we used only two samples for each writer for all the datasets created from the three databases used in this work.

2.2 Data Organization

Aiming to properly address the main investigations we intend to do in this work, we decided to organize the databases in several different versions. From now, we will refer to these versions by using the name of the database, followed by a subscripted text with the number of writers considered in the dataset, and the number of samples per writer, respectively. For example, the dataset created from the IAM database, considering only a single sample for all the 657 writers, will be referred to as $IAM_{657,1}$.

Considering that some writers have only one handwriting document available in some databases, we divided each document into nine blocks of size $(m \times n)$, where m and n are, respectively, width and height from the preprocessed document divided by three. In this way, we obtained more than one sample from these writers, which tends to improve the performance for identification, as we increase the number of samples for training. This zoning pre-processing was evaluated on all databases. Figure 2 illustrates the splitting of a document in blocks.

Firstly, we evaluated the IAM, CVL and BFL databases using two documents per writer (i.e. $IAM_{301,2}$, $CVL_{310,2}$ and $BFL_{315,2}$), even for those who have more than one sample in the database. From these documents, we took six blocks from the nine created blocks to compose the training set, and the others three blocks were used to make the test set. In case of experiments without DF, samples from the same document was used both in train and test set. When DF protocol is used, six blocks from one document were used for training and others three blocks from a second document for testing.

We have done it three times per document (three folds): using the first six blocks, the first three and last three blocks, and the last six blocks as training sets, respectively. The results described in Sect. 3 correspond to the mean and the standard deviation of these three executions.

In the second experiment, we intended to evaluate the performance using documents from all the 657 writers of the IAM dataset, but using DF only for those writers who have more than one manuscript sample. In case of writers from whom we have more than one document available (i.e. 301 writers), we also set three testing folds, but using blocks exclusively taken from the second document. Following the nomenclature notation defined here, this part of the data is called $IAM_{301,2}$. It is worth mentioning that, in case of single-sample writers (i.e. 356 writers), we kept the blocks from the same document, as it was the only option in order to use the all writers of the dataset. Considering the nomenclature defined here, this part of the data is called $IAM_{356,1}$. We have used the same classification model in both cases.

After that, we isolated the 356 single-sample writers and performed the same classification scheme to evaluate only the behaviour of the data, from which it is not possible to make investigations considering the DF protocol. These experiments were performed trying to get a better view of the influence of this part of the dataset on the results performed with the dataset as a whole. This subset is called $IAM_{356,1}$, as we are using strictly the single-sample writers.

To analyze the impact of DF, we also set a scenario in which the filter can be applied to every writer. For this purpose, we selected the first two documents from each writer with two or more documents (i.e. 301 writers). In this way, the DF protocol already described can be suitably applied to these data. The subset used here is $IAM_{301,2}$, as we took two documents from the writers with at least two samples. These versions created for IAM do not make sense for CVL and BFL databases, as they always have more than one sample for each writer.

2.3 Descriptors

Considering that texture is one of the main visual attributes on manuscript images, and also results previously described in the literature[8,16,23], in this work, we have used the following feature extractors: BSIF [10], EQP [15], LDN [19], oBIF [5], LETRIST [21], SURF [1]. In Table 1 are described the main parameters used by these descriptors. Details and codes can be found in their respective references.

Table 1. Features dimensions and main parameters.

Feature	Parameters	Dimensions
BSIF	*filter* = *ICAtextureFilters-11*×11-*8bit*	256
EQP	*loci* = *ellipse*	256
LDN	$mSize = 3$; $mask = kirsch$; $\sigma = 0.5$	56
LETRIST	$sigmaSet = 1, 2, 4$; $noNoise$	413
oBIF	$\alpha = 2, 4$; $\varepsilon = 0.001$	484
SURF	$SurfSize = 64$	257

3 Experimental Results and Discussion

Our main goal in this work is to evaluate if the neglection in the use of the DF protocol can lead to misleading results. For this, we choose the three aforementioned databases, and six different feature descriptors that have already proved to be efficient on the task investigated here.

Support Vector Machine (SVM) was chosen as the classifier because it has already been successfully used in this task in several works described in the literature [2,3,6,20]. The parameters C and γ from SVM classifiers were determined by grid search with $5 - fold$ cross-validation, using LibSVM [4] library. Once the SVM predictions are obtained individually for each block of the document, we perform a fusion of these predictions in order to get the final decision for the document as a whole. In this sense, we used the Sum Rule to combine the classifier's output [11], as it has already shown a good performance in this task [3].

The rates presented in this Section are, in many cases, lower than those already published in the literature. However, in this paper, we are not pursuing the improvement of the performance in terms of rates, but we aim to evaluate the impact when parts of the same document are used both on train and test sets. Furthermore, the reduction of performance is even expected to a certain extent, because hypothetically we are making the task investigated somehow more difficult.

The results described in Table 2 show us that regardless the texture descriptor used, there is a difference on performance when the DF protocol is employed. In this cases, the performance rates decrease if compared to the rates obtained without the use of DF for all texture descriptor evaluated. These results evidenced that the use of data taken from the same manuscript simultaneously both on train and test sets tends to positively bias the results. In the most extreme case, the difference between the rates obtained with or without the use of DF reaches 30.94% points, for the LDN descriptor. In the minority of cases, the difference is not significant if the standard deviation is observed (i.e. BSIF Descriptor). In addition, we can highlight the performance of the SURF descriptor as it presents the best rate and lowest standard deviation.

Table 2. Performance (%) using $IAM_{301,2}$ dataset

Descriptor	Without DF (σ)	With DF (σ)	DIFF*
BSIF	74.26 (±6.59)	70.95 (±4.88)	3.31
EQP	61.41 (±5.40)	54.87 (±3.69)	6.54
LDN	62.85 (±7.18)	31.91 (±2.57)	30.94
LETRIST	60.69 (±5.27)	31.84 (±2.55)	28.85
oBIF	80.40 (±6.47)	65.38 (±3.84)	15.02
SURF	84.89 (±3.29)	69.97 (±2.40)	14.92

* From this table, DIFF stands for the absolute difference between the rates with or without DF.

SURF, oBIF and BSIF continue the Top-3 best descriptors also on the CVL database, as we can see in Table 3. CVL database presented a high standard deviation regardless of the texture descriptor used. In this experiment, documents one and three of the CVL database were used, both written in English. We can also observe a difference between rates showed by strategies with and without DF. However, due to the high standard deviation presented, these differences may be statistically questionable.

Table 3. Performance (%) using $CVL_{310,2}$ dataset

Descriptor	Without DF (σ)	With DF (σ)	DIFF
BSIF	78.50 (±6.47)	67.17 (±4.19)	11.33
EQP	67.15 (±4.57)	55.86 (±5.28)	11.29
LDN	60.81 (±6.11)	48.40 (±6.47)	12.41
LETRIST	61.78 (±8.82)	45.72 (±4.56)	16.06
oBIF	85.05 (±7.04)	75.13 (±6.46)	9.92
SURF	79.30 (±9.26)	63.15 (±7.68)	16.15

The results achieved using the BFL database confirm that the SURF, oBIF, and BSIF are the best texture descriptors for the writer identification task when blocks are used (see Table 4). The oBIF texture descriptor presents a remarkable result, as it proved to be robust even when the DF protocol is applied.

Table 4. Performance (%) using $BFL_{315,2}$ dataset

Descriptor	Without DF (σ)	With DF (σ)	DIFF
BSIF	81.06 (±3.89)	74.91 (±3.23)	6.15
EQP	58.84 (±11.01)	45.72 (±3.86)	13.12
LDN	60.10 (±9.63)	42.10 (±3.53)	18.00
LETRIST	67.65 (±5.84)	46.92 (±4.21)	20.73
oBIF	88.04 (±3.09)	84.83 (±3.01)	3.21
SURF	89.06 (±3.18)	79.80 (±3.58)	9.26

As already pointed, taking into account similar conditions in other classification tasks investigated in the literature, the use of parts taken from the same document both in the train and test sets is supposed to introduce a bias in favour of the results.

However, some descriptors can contribute to more robust models so that they can generalize very well, even when the DF protocol is imposed. Results presented with the IAM and BFL databases reinforce this hypothesis, that descriptors which present lower rates are more impacted with the use of the DF protocol.

We performed a statistical test for the texture descriptor with the best performance (SURF), and for LETRIST, the descriptor with the biggest difference in results achieved with and without the DF protocol. We applied the Student's T-test, this statistical analysis shows us that there is statistical significance between results (P-value $\leq$ 5%) both on IAM and BFL databases. The lower was the P-value, we have stronger evidence against the null hypothesis. The P-values for IAM, CVL and BFL were $3.7e^{-09}$%, 0.032% and 0.001%. We have a null hypothesis on the CVL database, that is, we have enough evidence that the results are different from each other. The P-values were evaluated using the LETRIST descriptor. In this case, all rates were smaller with P-value, 7.8e^{-13}%, 0.0006% and 3.0e^{-07}% on IAM, CVL and BFL respectively.

Next, we evaluate the impact of the DF protocol using the SURF descriptor on all the IAM versions introduced in the Sect. 2.2 (i.e. $IAM_{657,1}$, $IAM_{356,1}$, and $IAM_{301,2}$). We will present the experiments in the same order that described the subsets in Sect. 2.2. In all experiments reported below were used the parameters previously described. In the first case, the $IAM_{657,1}$ database was used, such a way that all the 657 writers were considered. In this case, we did not use the DF protocol, as the $IAM_{657,1}$ is composed by a single document per writer. As we can see in Table 5, the identification rate achieved using all writers without DF was 70.12%.

In the following experiments, we check the writer identification performance on the IAM dataset, applying the DF protocol when it is possible, and not using the DF protocol for those writers with just one document. In this way, we conducted an evaluation using the subsets $IAM_{301,2}$ (with DF protocol) and $IAM_{356,1}$ (without DF protocol). We check the writer identification performance on the IAM dataset using the DF protocol only for writers to which it is possible. In this experiment, we obtained an identification rate of 63.83%. The results obtained suggest that the use of DF protocol tends to make the writer identification task harder than when it is not applied, as its usage has impacted the identification rate with a fall of 6.29%, and a P-value smaller than 5%. These results were not included in Table 5, as the DF protocol was applied in part of the database and in another part they are not applicable.

In the next experiment, we evaluate $IAM_{356,1}$ subset with 356 writers and one document per writer. In this case, the DF protocol was not used. The experiment performed in this scenario achieved a high identification rate, 81.83%. Lastly, we evaluate the $IAM_{301,2}$ subset with 301 writers and two document samples per writer. In this experiment, we can evaluate two scenarios, with the DF protocol and without the DF protocol. The first scenario was addressed for obvious reasons, and the second was performed aiming to compare the impacts of DF protocol on exactly the same subset. The results are described in Table 5. The P-value obtained using Student's t-test was 3.7e^{-09}%, that is, smaller than the threshold established on the literature (5%). Thus, we can conclude that there is a significant difference between applying or not the DF protocol.

As we can see in Table 5, the lowest identification rates happens when the DF protocol is employed. Moreover, the rate is 14.92% points lower than that obtained exactly on the same dataset but not using the DF protocol.

Table 5. Identification rates (%) using SURF descriptor

Training	# Writers	Without DF (σ)	With DF (σ)	DIFF
$IAM_{657,1}$	657	70.12 (±5.80)	–	–
$IAM_{356,1}$	356	81.83 (±1.06)	–	–
$IAM_{301,2}$	301	84.89 (±3.29)	69.97 (±2.40)	14.92

In addition, we show in Fig. 3 that the use of the DF protocol produces lower rates in all cases. It is one more piece of evidence that corroborates the rationale behind this work. It is still important to point out that although the performance rates are lower using the DF protocol, we believe these rates are more realistic.

4 Critical Review

At a first glance upon the problem addressed in this work, even those researchers used to the pattern recognition literature can think that this discussion is not applicable, since the use of different parts of the data taken from the same instance both in the training and test sets may be a controversial practice, out of place in several other application domains. However, it is important to observe that this kind of fine-tuning regarding the experimental protocol may vary widely from one application domain to another. As already mentioned here, Pampalk et al. [17] raised an intriguing question in 2005 about the results obtained until that moment in the music genre classification task. As a consequence, the authors introduced a new concept (i.e. artist filter) which has influenced most of the works in that field of research, having been considered as a restriction almost mandatory from that moment on.

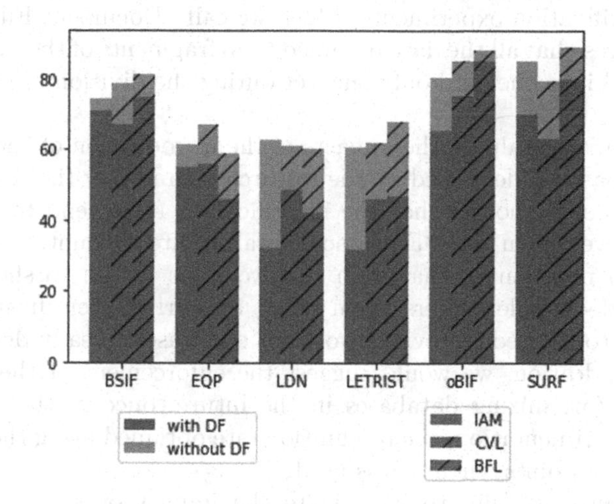

Fig. 3. Bar graph with results with or without DF.

In the case of the writer identification task, there are limits naturally imposed on the research community due to the limitations of many databases typically used in this kind of work. This is why it is not particularly rare to find works concerning writer identification in which data taken from the same manuscript are used both on the training and test sets, which is inconceivable in other application domains.

There are some factors that can help to understand why the lack of the DF protocol can somehow introduce a bias to the results on writer identification. That is because the same person has variations in their writing style on different occasions. In this regard, we can observe: the emotional state of the writer; the fluency of the pen/pencil used for writing; the lighting level around the writer; and the paper writability (smooth/rough), among others [13]. So, if we use fragments taken from the same manuscript both on train and test sets, the identification rates tend to be biased.

That said, we should consider the feasibility of the use of the DF protocol when creating new manuscript databases, avoiding single-sample writers, as it could introduce some bias to the identification rates obtained using those data.

Finally, the last important matter to cogitate here regards alternatives to minimize the impacts of single-sample writers on the development of experiments performed on databases already available in which it occurs. The researchers could consider performing experiments applying the DF protocol on those parts of the data in which it is possible, like done here.

5 Concluding Remarks

In this work, we proposed a novel restriction to be imposed for the development of writer identification experiments, which we call "Document Filter". The DF protocol enforces that all the data obtained from fragments of the same document must be placed into one, and only one, set during the division to create training and test sets.

In this vein, we evaluate the impacts of the introduction of the DF protocol on three databases widely used by the research community (i.e. IAM, CVL and BFL). Experiments showed that the identification rates tend to decrease at a considerable level when the DF protocol is taken into account.

It is worth mentioning that with this work we intend to show the effects of having single-sample writers when doing the writer identification task. We do not intend to discredit previous works or databases already described in the literature. In addition, we would suggest the enforcement of the DF protocol when creating/organizing databases in the future, once in the most extreme case of our experimentation, the recognition rate obtained using the DF protocol dropped 30.94% points when it was used.

In the future, we aim to investigate the impact of the availability of a restricted amount of text (i.e. only one paragraph per writer, only one line per writer, and only one word) on the performance of writer identification systems.

Acknowledgment. We thank the Brazilian research support agencies: Coordination for the Improvement of Higher Education Personnel (CAPES), and National Council for Scientific and Technological Development (CNPq) for their financial support.

References

1. Bay, H., Tuytelaars, T., Van Gool, L.: SURF: speeded up robust features. In: Leonardis, A., Bischof, H., Pinz, A. (eds.) ECCV 2006. LNCS, vol. 3951, pp. 404–417. Springer, Heidelberg (2006). https://doi.org/10.1007/11744023_32
2. Bertolini, D., Oliveira, L.S., Costa, Y.M.G., Helal, L.G.: Knowledge transfer for writer identification. In: Mendoza, M., Velastín, S. (eds.) CIARP 2017. LNCS, vol. 10657, pp. 102–110. Springer, Cham (2018). https://doi.org/10.1007/978-3-319-75193-1_13
3. Bertolini, D., Oliveira, L.S., Justino, E., Sabourin, R.: Texture-based descriptors for writer identification and verification. Expert Syst. Appl. **40**(6), 2069–2080 (2013)
4. Chang, C.C., Lin, C.J.: LIBSVM: a library for support vector machines. ACM Trans. Intell. Syst. Technol. **2**, 27:1–27:27 (2011)
5. Crosier, M., Griffin, L.D.: Using basic image features for texture classification. Int. J. Comput. Vis. **88**(3), 447–460 (2010)
6. Durou, A., Al-Maadeed, S., Aref, I., Bouridane, A., Elbendak, M.: A comparative study of machine learning approaches for handwriter identification. In: 2019 IEEE 12th International Conference on Global Security, Safety and Sustainability (ICGS3), pp. 206–212. IEEE (2019)
7. Freitas, C., Oliveira, L.S., Sabourin, R., Bortolozzi, F.: Brazilian forensic letter database. In: 11th International workshop on Frontiers on Handwriting Recognition, Montreal, Canada (2008)
8. Hannad, Y., Siddiqi, I., El Kettani, M.E.Y.: Writer identification using texture descriptors of handwritten fragments. Expert Syst. Appl. **47**, 14–22 (2016)
9. He, S., Schomaker, L.: Deep adaptive learning for writer identification based on single handwritten word images. Pattern Recogn. **88**, 64–74 (2019)
10. Kannala, J., Rahtu, E.: BSIF: binarized statistical image features. In: Proceedings of the 21st International Conference on Pattern Recognition (ICPR 2012), pp. 1363–1366 (2012)
11. Kittler, J., Hater, M., Duin, R.P.: Combining classifiers. In: Proceedings of 13th International Conference on Pattern Recognition, vol. 2, pp. 897–901. IEEE (1996)
12. Kleber, F., Fiel, S., Diem, M., Sablatnig, R.: CVL-database: an off-line database for writer retrieval, writer identification and word spotting. In: 2013 12th International Conference on Document Analysis and Recognition, pp. 560–564, August 2013
13. Koppenhaver, K.M.: Forensic Document Examination: Principles and Practice. Springer, Heidelberg (2007). https://doi.org/10.1007/978-1-59745-301-1
14. Marti, U.V., Bunke, H.: The IAM-database: an English sentence database for offline handwriting recognition. Int. J. Doc. Anal. Recogn. **5**, 39–46 (11 2002)
15. Nanni, L., Lumini, A., Brahnam, S.: Local binary patterns variants as texture descriptors for medical image analysis. Artif. Intell. Med. **49**(2), 117–125 (2010)
16. Newell, A.J., Griffin, L.D.: Writer identification using oriented basic image features and the delta encoding. Pattern Recogn. **47**(6), 2255–2265 (2014)
17. Pampalk, E., Flexer, A., Widmer, G., et al.: Improvements of audio-based music similarity and genre classificaton. In: ISMIR, London, UK, vol. 5, pp. 634–637 (2005)
18. Pekalska, E., Duin, R.P.: Dissimilarity representations allow for building good classifiers. Pattern Recogn. Lett. **23**(8), 943–956 (2002)
19. Ramirez Rivera, A., Rojas Castillo, J., Oksam Chae, O.: Local directional number pattern for face analysis: face and expression recognition. IEEE Trans. Image Process. **22**(5), 1740–1752 (2013)

20. Rehman, A., Naz, S., Razzak, M.I.: Writer identification using machine learning approaches: a comprehensive review. Multimedia Tools Appl. **78**(8), 10889–10931 (2018). https://doi.org/10.1007/s11042-018-6577-1
21. Song, T., Li, H., Meng, F., Wu, Q., Cai, J.: Letrist: locally encoded transform feature histogram for rotation-invariant texture classification. IEEE Trans. Circ. Syst. Video Technol. **28**(7), 1565–1579 (2018)
22. Wu, X., Tang, Y., Bu, W.: Offline text-independent writer identification based on scale invariant feature transform. IEEE Trans. Inf. Forensics Secur. **9**(3), 526–536 (2014). https://doi.org/10.1109/TIFS.2014.2301274
23. Xiong, Y., Wen, Y., Wang, P.S.P., Lu, Y.: Text-independent writer identification using sift descriptor and contour-directional feature. In: 2015 13th International Conference on Document Analysis and Recognition (ICDAR), pp. 91–95 (2015)

An Approach for BCI Using Motor Imagery Based on Wavelet Transform and Convolutional Neural Network

Lenka Rabčanová(✉) and Radoslav Vargic

FEI STU in Bratislava, 841 04 Bratislava, Slovakia
xrabcanova@stuba.sk

Abstract. In the presented contribution we propose a motor imagery based Brain-Computer Interface system for device control. Based on the EEG datasets we perform a two and three class classification of selected features for real motor and motor imagery movements. Features are created using complex wavelet transform. The classification is based on convolution neural networks. The results show that the method has a similar performance as the known reference method for given datasets.

Keywords: Brain-computer interface · EEG · Motor imagery · Wavelet transform · Convolutional neural network

1 Introduction

The aim of the Brain-Computer Interface (BCI) is to enable direct communication between humans and computers using control signals in the form of biological signals from the brain, which can be measured and interpreted by various methods.

The most common method of measurement of brain activity is a non-invasive method called Electroencephalography (EEG). It uses electrodes placed on the surface of the head as sensors that detect electrical changes in brain activity [1].

EEG based BCI is widely used in the combination with Motor Imagery (MI). MI is a cognitive process in which the subject imagines movement without any muscular activity. Sensorimotor rhythms (SMR) are generated in somatic, sensorimotor areas and are concentrated mainly in the Alpha (8–13 Hz), Beta (13–30 Hz) and Gamma (more than 31 Hz) frequency bands. Execution or imagination of movements of individual parts of the body, e.g. legs, creates a unique response in the SMR [5, 6].

Nowadays there are still a lot of obstacles to using EEG based BCI. In general, the signal has poor spatial resolution and a low signal-to-noise ratio (SNR). During signal measurement, various artifacts and interferences are mixed with the desired information signal. Undesirable factors that affect the signal can be, for example, eye blinking, muscle activity, and background activities during signal acquisition [7, 8]. Using the BCI system along with the MI can be implemented in various fields and has a wide range of uses, such as drone control [15] (in particular the unmanned aerial vehicles), robotic arm [16],

G. Rozinaj and R. Vargic (Eds.): IWSSIP 2021, CCIS 1527, pp. 185–197, 2022.
https://doi.org/10.1007/978-3-030-96878-6_17

various robots [17], as additional modality in Human Computer Interaction (HCI), e.g. to replace the gestures [19] or to complement saliency information [20] when sensing a scene.

For MI-based BCI research, there are available more datasets. In the following summary, the authors used the Gelwin Schalk Dataset (2009) [9, 10].

The authors in [11] selected different EEG channels (especially in the frontal lobe area) and compared their relevance using Fast Fourier Transform (FFT) to extract the features such as mean, standard deviation and skewness in spectrum split into bands. The real movement data were used for training and the MI data were subsequently used as a test dataset. Their goal was to differentiate the real movement from the same type of MI movement while finding a correlation between these. K-nearest neighbors algorithm (KNN) was used as a classification method obtaining accuracy 48.2%–52.8% (on the different EEG channels). With careful selection of relevant features, they were able to obtain accuracy from 46.3% to 79.8%. The best results were obtained from the electrodes Fc1, Fcz, Fc2, F6 and Ft8.

In [13] the authors used a convolutional neural network for extracting and classifying EEG signals. Their model reached an average cross-validation accuracy of 87.98%, 76.61%, 65.73% respectively, for the two-, three-, and four-class classification tasks. After they reduced the amount of input data to the first three seconds, global accuracy values obtained were 80.38%, 69.82% and 58.58% respectively.

The authors in [12] performed two-, three-, and four-class classification:

- 2-class: MI opening/closing left or right fist,
- 3-class: MI opening/closing left or right fist and Baseline data,
- 4-class: MI opening/closing left or right fist, Baseline, MI opening/closing both feet.

They used Qnet, as feature extraction and classification method. As they claimed in the paper, Qnet is a Neural Network designed to differentiate EEG signal, and the importance of different electrodes data, and in addition to extracting relevant features. Using 80% as training data and 20% as testing data from all 64 electrodes, they achieved an accuracy of 82.88%, 74.75% and 65.82% in the two-, three- and four-class of classification, respectively. Authors compared their results to results described in [13] and they considered the results from [13] slightly worse.

Authors in [2] used dataset from BCI competition III: dataset IV-A [14]. They performed two-class classification of the right-foot and right-hand movements. In the feature extraction phase, the Short Time Fourier transform, and Continuous Wavelet transform were used. They obtained the best classification accuracy of 99.35% with the CNN called AlexNet.

In [3] the authors used the dataset from Technical University in Graz [4]. The goal was to differentiate right and left MI movements. They used different classification algorithms including LDA (Linear Discriminant Analysis), QDA (Quadratic classifier), GMM (Gaussian mixture model) and APNN (Adaptive probabilistic neural network). The classification accuracy obtained with these methods was 74.91%, 83.54%, 87.63% and 90.16% respectively. With the proposed APNN method they achieved greater accuracy than was the best result obtained in the BCI 2003 Competition, for which this dataset was originally measured. The highest value of classification accuracy achieved in this competition was 89.3%.

In this paper, we worked with EEG Gelwin Schalk Dataset (2009) [9, 10] performing two- and three-class classifications of the selected features, both for Real and MI movements. Features were created using Fast Fourier Transform (FFT), Short Time Fourier Transform (STFT) and Continuous wavelet transform (CWT). We tried different approaches and combinations of the mentioned methods to observe the impact on the classification results. As the classification methods, we used the KNN algorithm (as basic reference) and convolution neural network called AlexNet. Our goal was to differentiate most effectively MI movement from the state without movement (calm state).

In the Sect. 2, we explained data. In the Sect. 3, we proposed different methods. In the Sect. 4, we described results that we obtained using different methods and approaches. And in the last section, we discussed further possibilities for extending the proposed approaches.

2 Reference Dataset

As a reference dataset we decided to use the EEG Motor Movement/Imagery dataset [9, 10] as for this dataset there were available some reference methods to which we could compare [11–13]. This dataset contained measurements obtained from 109 subjects. Each of them performed 14 runs. Runs 1 and 2 were one-minute baseline runs (with open and closed eyes). Runs 3 to 14, were two-minute runs with real and MI movement. In runs 3 to 14, the subject has performed 4 different tasks (such as open and close left or right fist, imagine opening and closing left or right fist). In total there were 30 executions (subtasks) in each run. Data were sampled at 160 samples per second.

These tasks were performed while 64-channels (EEG 10-10 system) (Fig. 1) were recorded using the BCI2000 system. Measurements were obtained on 64 electrodes. During the following experiments, we mainly worked with electrodes C3, Cz and C4, as they are located in the sensorimotor center which is known to be responsible for activity and movement.

For our purposes, we chose data from runs 3 to 14, which were runs where the subjects performed left or right hand movements (both MI and Real movements). In particular we chose the following subtasks:

1. RF: opening and closing right fist (Real or MI movement
2. LF: opening and closing left fist (Real or MI movement)
3. Calm: rest period between RF and LF movements

In addition, we selected only data from the first four seconds of each subtask.

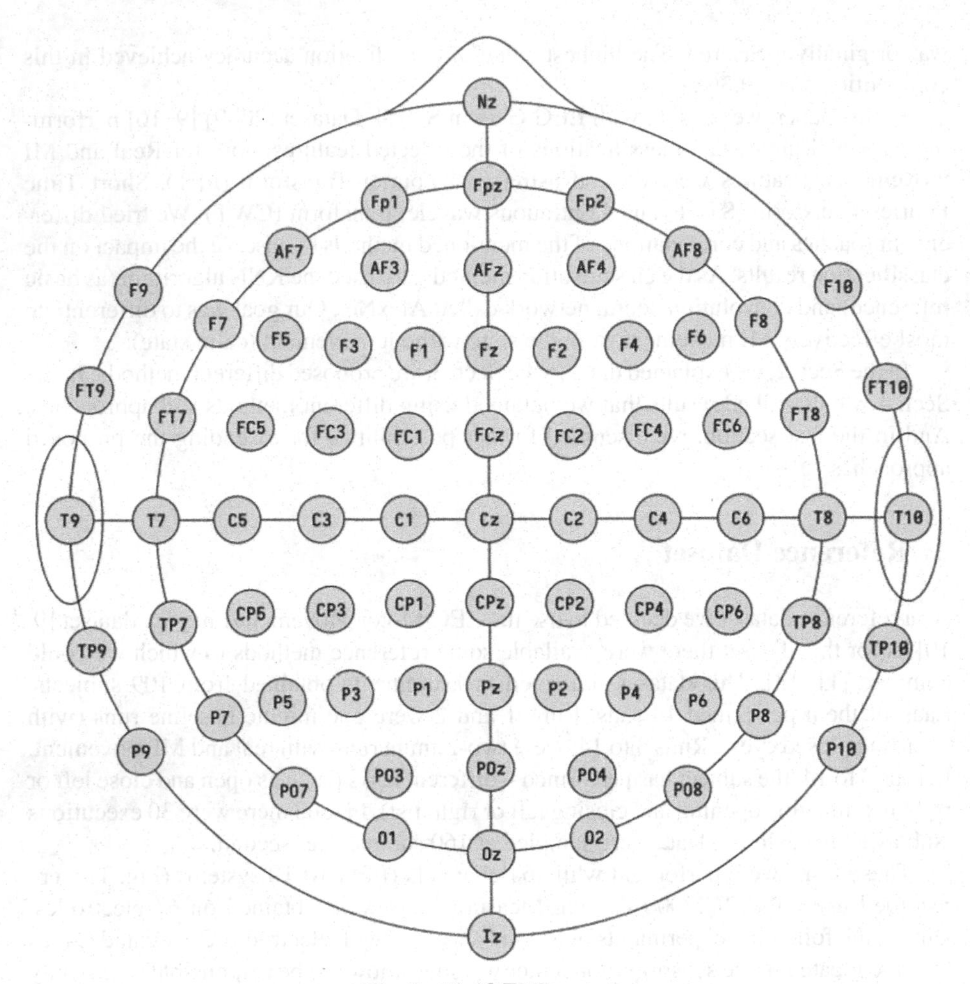

Fig. 1. 10-10 EEG system

3 Proposed BCI Realization

In this contribution, we used various methods, some of them only for basic performance comparison.

In the method, which we refer to as method A, we used frequency-based feature extraction coupled with a classical classifier. Feature extraction was performed by calculation of total energy in Alpha (8–13 Hz), Beta (13–30 Hz) and Gamma frequency (31–40 Hz) bands and their combinations. These bands are known for reflecting brain activity during MI and real movements. The total energy was calculated as a sum of the energy on all three of the mentioned electrodes. We used the k-nearest neighbors algorithm (KNN) as a classifier. For a better understanding of spectral energy feature relevance, we provide an example in Fig. 2, where we depict the KNN (k = 11) classification accuracy for pairs features denoted as F1 and F2 which measure the amount of energy within 5 Hz bandwidth (BW) around the specified frequency. Note the distorted band around 60 Hz (blue vertical and horizontal stripes forming a cross), which corresponds to AC energy distribution frequency.

In the method that we refer to as B, we used time-frequency based feature extraction. For this purpose, we selected spectrogram and scalogram as both belong to quadratic time-frequency distributions and show the distribution of energy in time and frequency. The spectrograms were computed using Short Time Fourier Transform (STFT) with various window functions and variable window size. The scalogram was computed using various Continuous Wavelet Transforms (CWT) such as Amor, Morse, Bump. As the time-frequency representation is two-dimensional, it is naturally suitable for image-based classification method such as Convolutional Neural Network (CNN).

For our experiments, we selected AlexNet [18]. AlexNet takes as input, images of size 227 × 227 × 3. As the AlexNet takes input images that have three color channels, we decided to use 2 types of inputs: single electrode and three electrodes.

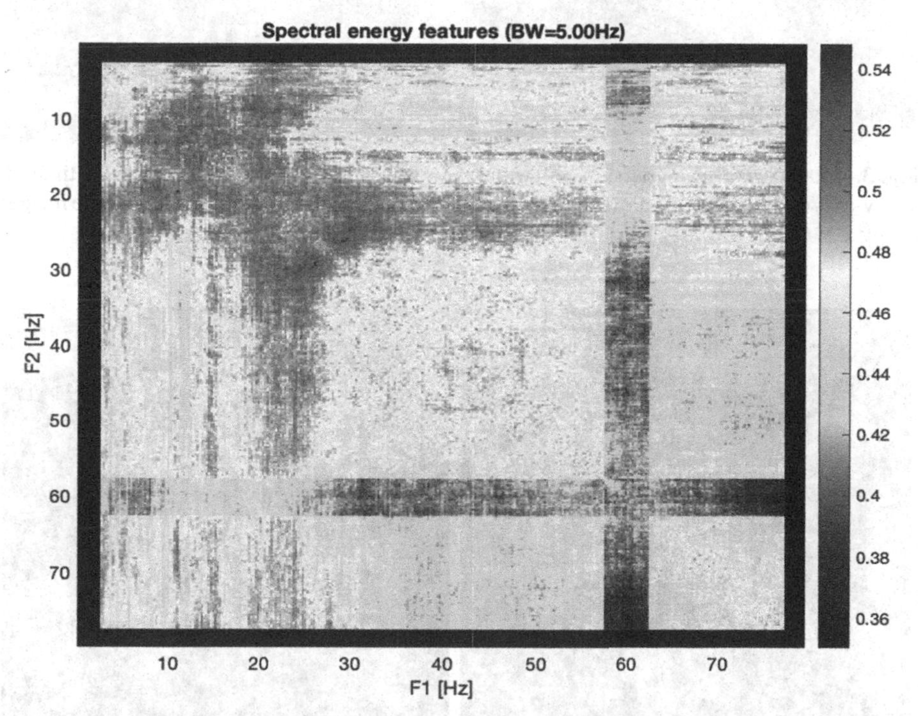

Fig. 2. Frequency energy relevance in 2D, measured using the KNN classifier (k = 11). The same size of training and testing data group (real movement data).

In a single electrode case, we have created grayscale images – the same values were mapped to each color channel. In the three-electrode case, we created RGB images, and we mapped each electrode to separate color channel (R-C3, G-Cz, B-C4). Corresponding example inputs for scalogram and spectrogram are shown in Figs. 3 and 4. These inputs cover the whole available frequency band from 0 Hz to 80 Hz (as the input sample rate was 160 Hz). The variants of method B include usage only of partial sub-bands (from 5 Hz to 30 Hz) which include Alpha and Beta sub-bands as they were considered as the most relevant based on experiments such as depicted in Fig. 2. Example scalograms and spectrograms are shown in Fig. 5.

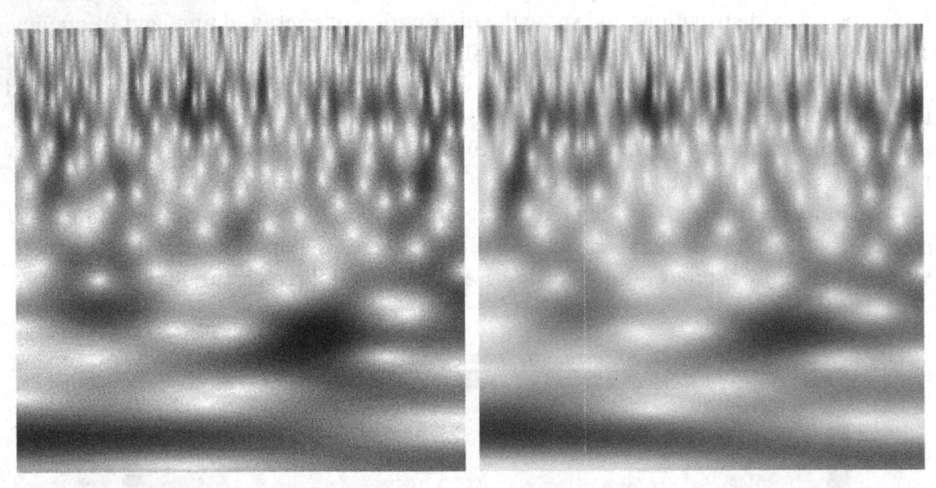

Fig. 3. Examples of scalogram images (with Amor wavelet). The x-axis corresponds to time (0–4 s), y-axis corresponds to scale (0–80 Hz equivalent). Left: single C3 electrode (grayscale image). Right: C3, Cz, C4 electrodes (RGB image). Real movement data.

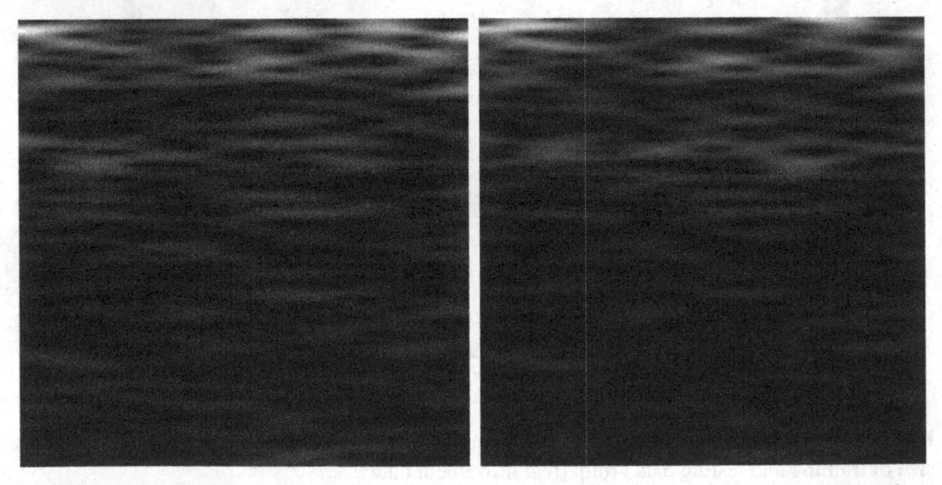

Fig. 4. Examples of spectrogram images (with 1s Hanning window). The x-axis corresponds to time (0–4 s), y-axis corresponds to frequency (0–80 Hz). Left: single C3 electrode (grayscale image). Right: C3, Cz, C4 electrodes (RGB image). Real movement data.

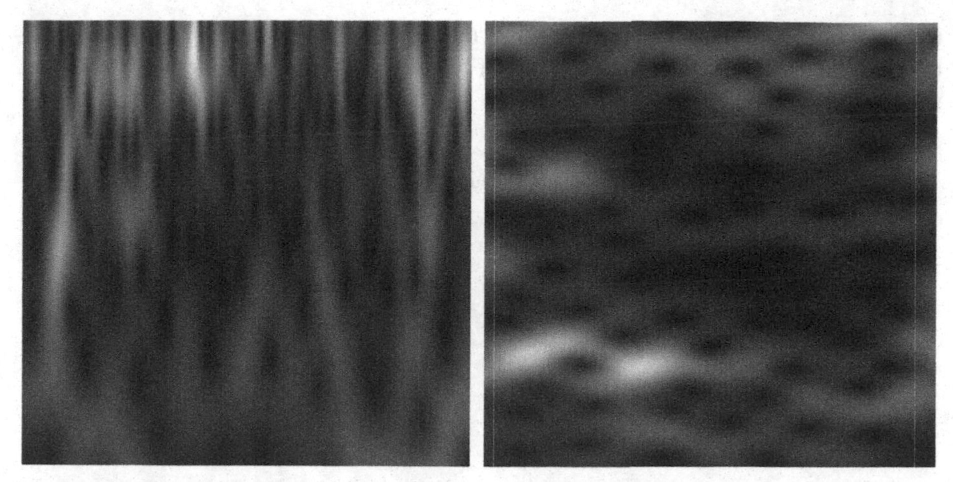

Fig. 5. Left: example of band-limited (5–30 Hz) scalogram. Right: example of band-limited (5–30 Hz) spectrogram. Based on the data depicted in Figs. 3 and 4.

4 Evaluation Results – Method A

With method A we have performed various experiments with configuration for two-class classifications, real and MI movement with opening and closing left/right fist.

Various combinations of Alpha, Beta and Gamma bands were explored, as well as a full search of variable-sized bands and their combinations. Table 1 is showing the best KNN classification results using Euclidian distance based on a different number of k-neighbors for real and MI movement. The data were used only from the first 55 subjects to speed up the analysis. The accuracy obtained was up to 59.1% performing real movement classification between left fist movement and calm state. And up to 58.8% performing MI movement classification between left fist and right fist movements.

Omitting the Gamma frequency band from the sum of the total energy had more impact on MI movement than it had on real movement, but overall, this impact was negligible.

The local optimum k-neighbors number varied across all the experiments, but overall results are showing that a smaller number of k is- performing better for the real movement classification while MI movement shows higher accuracy with a greater number of k. For a better understanding of how the frequency bands can be useful, we run a lot of tests with different hyperparameters with full frequency band scan using various selected bandwidths used for energy computation.

One of such configurations is shown as an example in Fig. 2, where the full scan search for the best combination of 2 independent frequency bands (frequency energy relevance in 2D) with bandwidth 5 Hz was used, and the accuracy of the resulting KNN classification was displayed. The number of k-neighbors was 11 and the classification was performed between right fist and calm state (real movement) for all 109 subjects. As one can see, the most important areas are around Alpha (8–15 Hz) and Beta (16–31 Hz), which confirms the theoretical background of this topic. On the opposite side, the bands that include 60 Hz frequency are significantly impaired what we attribute to the frequency of electricity distribution in the public grid. By varying the bandwidth, the same approximate importance layout of bands was discovered. For more examples, see Fig. 6.

Table 1. KNN best classification results for 2class real and MI movement.

Type	Classes	k-neighbors	Frequency bands	Accuracy
Real movement	LF, Calm	23	Alpha, Beta	59.1%
Real movement	LF, Calm	7	Alpha, Beta, Gamma	59.0%
MI movement	LF, RF	399, 403	Alpha, Beta	58.8%
MI movement	LF, RF	451	Alpha, Beta, Gamma	58.2%

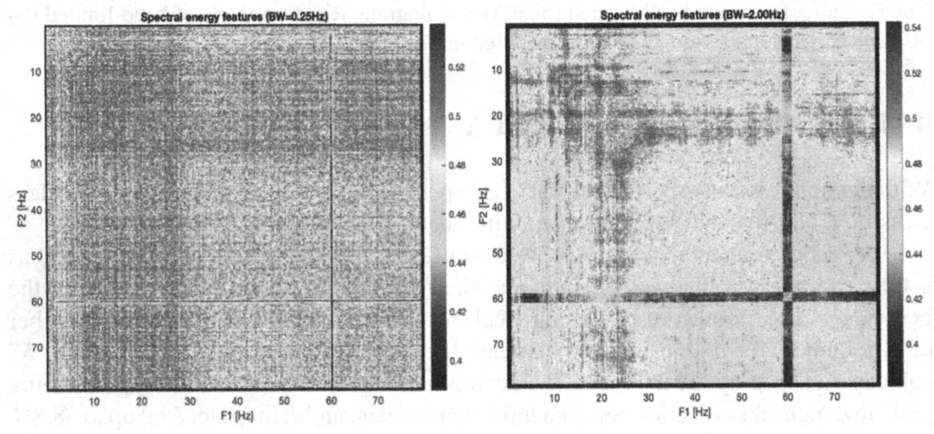

Fig. 6. Examples of frequency energy relevance with different bandwidths (0.25 Hz and 2 Hz), measured using the KNN classifier (k = 11). The same size of training and testing data group (real movement data).

5 Evaluation Results – Method B

We realized many 2-class and 3-class classification experiments using method B and its variants. The variants included the whole frequency band, the partial frequency band, scalograms (with various windowing functions and with variable window widths), spectrograms (with various wavelets), single-electrode resulting in grayscale or three-electrodes resulting in RGB image mappings.

In the beginning, we tuned the spectrogram time-frequency features. The performance differences when using Gauss, Blackman, Hanning and Hann windowing functions were negligible, more differences were found by using different windows sizes. Representative results using the Hanning window are shown in Table 2. For further comparisons, we have used the Hanning window and 1s window width.

We also have checked the relevance of the C3, Cz and C4 electrode selection, when using a three-electrode color mapping. We tried to change these electrodes to other electrodes (single change at a time) and checked, how it influences the results. Sample results when exchanging the electrode C4 are shown in Table 3. We see that electrode change to the nearby electrodes only led to performance degradation. We obtained similar results exchanging the electrodes C3 and Cz.

Table 2. AlexNet classification results for 2class real and MI movements for STFT using different size of Hanning window.

Type	Classes	Window size	Electrodes	Frequency	Accuracy
Real movement	Calm, LF	0.25 s	C3, Cz, C4	0–80 Hz	79.7%
Real movement	Calm, LF	0.5 s	C3, Cz, C4	0–80 Hz	76.6%
Real movement	Calm, RF	1 s	C3, Cz, C4	0–80 Hz	76.6%
Real movement	Calm, RF	0.125 s	C3, Cz, C4	0–80 Hz	75.5%
Real movement	Calm, LF	1.5 s	C3, Cz, C4	0–80 Hz	68.5%
Real movement	Calm, RF	0.25 s	C3, Cz, C4	5–30 Hz	78.1%
Real movement	Calm, LF	0.125 s	C3, Cz, C4	5–30 Hz	76.0%
Real movement	Calm, LF	0.5 s	C3, Cz, C4	5–30 Hz	74.4%
Real movement	Calm, LF	1 s	C3, Cz, C4	5–30 Hz	74.1%
Real movement	Calm, LF	1.5 s	C3, Cz, C4	5–30 Hz	71.5%

Table 3. AlexNet classification results for 2class real and MI movement (0–80 Hz) using measurements from different electrodes.

Type	Classes	Method	Electrodes	Accuracy
Real movement	Calm, RF	CWT (amor)	C3, Cz, C4	80.9%
Real movement	Calm, RF	CWT (amor)	C3, Cz, C5	73.5%
Real movement	Calm, RF	CWT (amor)	C3, Cz, C1	75.1%
Real movement	Calm, RF	CWT (amor)	C3, Cz, C2	73.6%
Real movement	Calm, RF	CWT (amor)	C3, Cz, C6	73.4%
MI movement	Calm, LF	CWT (amor)	C3, Cz, C4	74.3%
MI movement	Calm, LF	CWT (amor)	C3, Cz, C5	65.7%
MI movement	Calm, LF	CWT (amor)	C3, Cz, C1	66.5%
MI movement	Calm, LF	CWT (amor)	C3, Cz, C2	67.8%
MI movement	Calm, LF	CWT (amor)	C3, Cz, C6	67.7%

The overall best results for real and MI movement are shown in Table 4. The accuracy for the real movement was up to 80.9% while the accuracy for MI was up to 74.3%. Both results were obtained in 2class classification between left fist/right fist movement and calm state. In addition, the best accuracy was obtained using measurements from 3 electrodes at once (RGB image). When comparing accuracy from the point of view of different types of wavelets, higher accuracy is observed while using Amor. Method B variants that use the scalogram generated using Continuous Wavelet transform have significantly better results than variants using Short Time Fourier transform. The accuracy of the variants using STFT was up to 76.6% for the real movement and up to 65% for MI movement.

Table 4. AlexNet classification results for 2class real and MI movement (0–80 Hz).

Type	Classes	Method	Electrodes	Accuracy
Real movement	Calm, RF	CWT (amor)	C3, Cz, C4	80.9%
Real movement	Calm, RF	CWT (bump)	C3, Cz, C4	80.2%
Real movement	Calm, RF	CWT (morse)	C3, Cz, C4	79.4%
MI movement	Calm, LF	CWT (amor)	C3, Cz, C4	74.3%
MI movement	Calm, LF	CWT (morse)	C3, Cz, C4	70.5%
MI movement	Calm, LF	CWT (bump)	C3, Cz, C4	67.0%
Real movement	Calm, RF	STFT	C3, Cz, C4	76.6%
Real movement	Calm, LF	STFT	C3, Cz, C4	73.0%
MI movement	Calm, LF	STFT	C3, Cz, C4	65.0%
MI movement	Calm, RF	STFT	C3, Cz, C4	64.3%
Real movement	Calm, RF	CWT (amor)	C3	70.5%
Real movement	Calm, LF	CWT (amor)	C4	70.3%
Real movement	Calm, RF	CWT (amor)	Cz	66.5%
MI movement	Calm, RF	CWT (bump)	C3	64.3%
MI movement	Calm, RF	STFT	C4	64.3%
MI movement	Calm, RF	STFT	Cz	65.2%

Although the best results were obtained for three electrodes variants, it is worth mentioning also results for single electrode variants. The real movement accuracy on electrode C3 was up to 70.5% while classifying between calm state and right fist movement and on electrode C4 up to 70.3% classifying between calm state and left fist movement. The MI movement accuracy on electrodes C3 and C4 were up to 64.3% but in contrast with the real movement, the higher accuracy was observed while using bump wavelet along with STFT while classifying right fist movement and calm state. As assumed, the results obtained using the Cz electrode achieved lower accuracy (as this electrode is used as a reference electrode while measuring the EEG signal).

We also realized various experiments using classification between all three classes, but the higher accuracy observed was up to 57.80% for real movement and below 50% for MI movement. Altogether, AlexNet can distinguish better between one type of movement and calm state, than between two types of movements (or between more than 2 classes of data).

Based on the expectations, that the Gamma band could be redundant for the classification we tried to band limit the scalograms and spectrograms from 5 Hz to 30 Hz frequency band. The corresponding results related to the cases stated in Table 2 are listed in Table 5. As one can see, the achieved accuracy is lower, and the average difference is about 4.5%.

Table 5. AlexNet classification results for 2class real and MI movement (5–30 Hz).

Type	Classes	Method	Electrodes	Accuracy
Real movement	Calm, RF	CWT (amor)	C3, Cz, C4	79.9%
Real movement	Calm, RF	CWT (bump)	C3, Cz, C4	77.1%
Real movement	Calm, RF	CWT (morse)	C3, Cz, C4	79.9%
MI movement	Calm, LF	CWT (amor)	C3, Cz, C4	67.2%
MI movement	Calm, LF	CWT (morse)	C3, Cz, C4	59.3%
MI movement	Calm, LF	CWT (bump)	C3, Cz, C4	61.3%
Real movement	Calm, RF	STFT	C3, Cz, C4	71.5%
Real movement	Calm, LF	STFT	C3, Cz, C4	74.1%
MI movement	Calm, LF	STFT	C3, Cz, C4	61.3%
MI movement	Calm, RF	STFT	C3, Cz, C4	56.8%
Real movement	Calm, RF	CWT (amor)	C3	66.0%
Real movement	Calm, LF	CWT (amor)	C4	67.5%
Real movement	Calm, RF	CWT (amor)	Cz	67.4%
MI movement	Calm, RF	CWT (bump)	C3	59.3%
MI movement	Calm, RF	STFT	C4	52.2%
MI movement	Calm, RF	STFT	Cz	56.8%

To pursue the reason for this performance degradation, we formulated the hypothesis, that CNN could train to artefacts around 60 Hz, above 60 Hz and near DC components (signal drifts) which were wiped out after the band limitation. So, we constructed two other band limitations: from 1 Hz to 58 Hz and from 1 Hz to 80 Hz to be able to compare the results. The representative results using best cases for real and MI movement results are shown in Table 6. We can see that there was no abrupt performance degradation of these new cases against the unlimited 0–80 Hz case.

So, we concluded that the hypothesis was not valid, and the degraded performance of the 5–30 Hz case was caused by multiple factors, whose determination is not trivial nor straightforward.

Table 6. AlexNet classification results for 2class real and MI movement (1–80 Hz and 1–58 Hz).

Type	Classes	Method	Frequencies	Electrodes	Accuracy
Real movement	Calm, RF	CWT (amor)	1–58 Hz	C3, Cz, C4	81.9%
Real movement	Calm, RF	CWT (amor)	1–80 Hz	C3, Cz, C4	80.1%
Real movement	Calm, RF	CWT (amor)	0–80 Hz	C3, Cz, C4	80.9%
MI movement	Calm, LF	CWT (amor)	1–58 Hz	C3, Cz, C4z	70.2%
MI movement	Calm, LF	CWT (amor)	1–80 Hz	C3, Cz, C4	71.8%
MI movement	Calm, LF	CWT (amor)	0–80 Hz	C3, Cz, C4	74.3%

6 Conclusion

In the presented paper we realized multiple methods of motor imagery based Brain-computer interface system for device control.

Method A using simple frequency-based features achieved accuracy up to 59%. Method B using complex time-frequency feature extraction coupled with convolutional neural network achieved significantly better accuracy, up to 82%, which is comparable with results published in papers [12] and [13], proving the selected approach as competitive.

For the feature work, we would like to extend time-frequency features based on Modes such as Empirical Mode Decomposition and Variational Mode Decomposition. We would like to work on more sophisticated classification methods.

Acknowledgements. The research in this paper has been supported by VEGA 1/0440/19, the 2020-1-CZ01-KA226-VET-094346 DiT4LL ERASMUS+ Innovation Project and the Excellent creative team project VirTel.

References

1. Allison, B.Z., Krusienski, D.: Noninvasive brain-computer interfaces. In: Jaeger, D., Jung, R. (eds.) Encyclopedia of Computational Neuroscience. Springer, New York, NY (2014). https://doi.org/10.1007/978-1-4614-7320-6_707-1
2. Chaudhary, S., Taran, S., Bajaj, V., Sengur, A.: Convolutional neural network based approach towards motor imagery tasks EEG signals classification. IEEE Sens. J. 1–1 (2019). https://doi.org/10.1109/jsen.2019.2899645
3. Hazratti, M., Erfanian, A.: An online EEG-based brain–computer interface for controlling hand grasp using an adaptive probabilistic neural network. Med. Eng. Phys. **32**(7), 730–739 (2010)
4. Data set provided by Department of Medical Informatics, Institute for Biomedical Engineering, University of Technology Graz. (Gert Pfurtscheller) Correspondence to Alois Schlögl alois.schloegl@tugraz.at. http://bbci.de/competition/ii/index.html#datasets
5. Filho, S.S.C., Attux, R., Castellano, G.: Can graph metrics be used for EEG-BCIs based on hand motor imagery? Biomed. Signal Process. Control **40**, 359–365 (2018)

6. Tariq, M., et al.: Mu-beta rhythm ERD/ERS quantification for foot motor execution and imagery tasks in BCI applications. In: 2017 8th IEEE International Conference on Cognitive Infocommunications (CogInfoCom). IEEE, pp. 000091–000096 (2017)
7. Vaid, S., Singh, P., Kaur., C.: EEG signal analysis for BCI interface: a review. In: 2015 Fifth International Conference on Advanced Computing & Communication Technologies, IEEE, pp. 143–147 (2015)
8. Schomer, D.L.: The normal EEG in an adult. In: Blum, A.S., Rutkove, S.B. (eds.) The Clinical Neurophysiology Primer. Humana Press, Totowa, NJ (2007). https://doi.org/10.1007/978-1-59745-271-7_5
9. Schalk, G., McFarland, D.J., Hinterberger, T., Birbaumer, N., Wolpaw, J.R.: BCI2000: A general-purpose brain-computer interface (BCI) system. IEEE Trans. Biomed. Eng. **51**(6), 1034–1043 (2004)
10. Goldberger, A., Amaral, L., Glass, L., Hausdorff, J., Ivanov, P.C., Mark, R., Stanley, H.E.: PhysioBank, PhysioToolkit, and PhysioNet: components of a new research resource for complex physiologic signals. Circulation [Online] **101**(23), e215–e220
11. Alpturk, E.K., Kutlu, Y.: Analysis of relation between motor activity and imaginary EEG records. arXiv preprint arXiv:2101.10215 (2021)
12. Fan, C.C., et al.: Bilinear neural network with 3-D attention for brain decoding of motor imagery movements from the human EEG. Cogn. Neurodyn. 15(1), 181–189 (2021). https://doi.org/10.1007/s11571-020-09649-8
13. Dose, H., et al.: An end-to-end deep learning approach to MI-EEG signal classification for BCIs. Expert Syst. Appl. **114**, 532–542 (2018)
14. Data set provided provided by the Berlin BCI group: Fraunhofer FIRST, Intelligent Data Analysis Group (Klaus-Robert Müller, Benjamin Blankertz), and Campus Benjamin Franklin of the Charité - University Medicine Berlin, Department of Neurology, Neurophysics Group (Gabriel Curio), Correspondence to Benjamin Blankertz ⟨benjamin.blankertz@tu-berlin.de⟩. http://www.bbci.de/competition/iii/desc_IVa.html
15. Choi, J.W., Kim, B.H., Jo, S.: Asynchronous motor imagery brain-computer interface for simulated drone control. In: 2021 9th International Winter Conference on Brain-Computer Interface (BCI), IEEE, pp. 1–5 (2021)
16. Mwata-Velu, T., et al.: Motor imagery classification based on a recurrent-convolutional architecture to control a hexapod robot. Mathematics **9**(6), 606 (2021)
17. Zhang, J., Wang, M.: A survey on robots controlled by motor imagery brain-computer interfaces. Cogn. Robot. **1**, 12–24 (2021)
18. Krizhevsky, A., Sutskever, I., Hinton, G.E.: ImageNet classification with deep convolutional neural networks. In: Advances in Neural Information Processing Systems (2012). https://proceedings.neurips.cc/paper/2012/file/c399862d3b9d6b76c8436e924a68c45b-Paper.pdf
19. Vančo, M., Minárik, I., Rozinaj, G.: Gesture identification for system navigation in 3D scene. In: 54th ELMAR International Symposium ELMAR-2012, ELMAR Proceedings, pp. 45–48, September 12–14, Zadar, Croatia, ISSN: 1334-2630, ISBN: 978-953-7044-13-8, WOS:000399723300010 (2012)
20. Polakovič, A., Vargic, R., Rozinaj, G.: Adaptive multimedia content delivery in 5G networks using DASH and saliency information. In: 25th International Conference on Systems, Signals and Image Processing (IWSSIP), June 20–22, Maribor, Slovenia, ISSN: 2157-8672, ISBN: 978-1-5386-6979-2, WOS: 000451277200008 (2018). https://doi.org/10.1109/IWSSIP.2018.8439215

Advanced Scene Sensing for Virtual Teleconference

Ivan Minárik(✉), Marek Vančo, and Gregor Rozinaj

Slovak University of Technology, 841 04 Bratislava, Slovakia
{ivan.minarik,marek_vanco,gregor.rozinaj}@stuba.sk

Abstract. Over the last year the need for video conferences has risen significantly due to the ongoing global pandemic. The goal of this project is to improve user experience from having access to only voice and plain 2D image by adding a third spatial dimension, creating a more immersive setting. Azure Kinect Development Kit utilizes multiple cameras, namely the RGB camera and depth camera. Depth camera is based on ToF principle, which uses near-IR to cast modulated illumination onto the scene. The setup uses multiple Azure Kinect devices in sync and offset in space to obtain non-static 3D capture of a person. Unity engine with Azure Kinect SDK is used to process the data gathered by all devices. Firstly, a depth spatial map is created by combining overlaid outputs from each device. Secondly, RGB pixels are mapped onto depth spatial points to provide a final texture to the 3D model. Taking into account the need to export a continuous capture of raw data to a server, body tracking and image processing algorithms are used. Finally, the processed data can be exported and utilized in AR, VR or any other 3D capable interface. This 3D projection aims to enhance sensory experience by utilising non-verbal communication along with classical speech in video conferences.

Keywords: 3D sensing · Volumetric alignment · Point cloud

1 Introduction

Over the last year the need for video conferences has risen significantly due to the ongoing global pandemic. More and more people stay home, whole countries close borders and many people work remotely. Personal, work or school meetings have migrated to the online world where people are forced to use currently available communication technology like Google Meet, Zoom or Microsoft Teams.

These communication technologies have multiple useful features like support for a large number of connected users in real time, support for sending multimedia and some services serve as cloud storage for groups of people and all of that with high reliability and simplicity. But even though with all of these benefits, virtual meeting is still incomparable with real life meeting. There are many reasons, one of which states that verbal and vocal communication contributes to only 45% of interpersonal communication. The rest is composed of

G. Rozinaj and R. Vargic (Eds.): IWSSIP 2021, CCIS 1527, pp. 198–209, 2022.
https://doi.org/10.1007/978-3-030-96878-6_18

non-verbal communication. This kind of communication includes body posture, distance between people while talking, gesticulation, facial expressions and many more. Such aspects of communication are not even captured because of current limitations of our technology. Also as a next major reason could be lack of a third special dimension which current web cameras are not capable of capturing and transmitting over the network [1].

Authors' research team published several papers in the topic on experience-based videoconferencing, which is a new generation of standard videoconferencing approach [2–4]. In this paper we would like to focus on these problems (address these shortcomings) and how to solve them. We came up with a solution which could enrich such meetings with all of the mentioned missing aspects of communication.

2 Depth-Sensing Approaches

The concept of depth sensing can be described as capturing spatial and RGB aspects of a scenery. The three-dimensional depth image combined with the 2D RGB image result in a real time 3D reconstruction of a scene. Such technology has a broad spectrum of uses in many different fields. For example, robotics, medicine, telecommunication, architecture, etc. There are plenty of devices capable of depth sensing on the market. However, they differ in the technology that is used to capture and process the depth information. Common are RGB cameras and infrared sensors. Some devices use multiple RGB cameras to create the depth map. That is particularly demanding on the computation, but produces a higher resolution depth map. On the other hand, devices that use infrared sensors in addition to RGB cameras lack the high-resolution in-depth map, but are less demanding on the system. Therefore, the choice of the device depends on the environment in which it is going to be used.

2.1 Depth Sensing Using Structured Light

The principle of using structured light (SL) lies in active stereo-vision method. SL uses a preconfigured sequence of light in a geometric pattern, which is periodically projected onto an object. The pattern, originally in 2D plane, deforms along the shape of an object tracing its structure as it can be seen in Fig. 1. This deformed pattern is then captured by a camera from a different angle, analysed and processed creating a depth map [5].

2.2 Depth Sensing Using Time-of-Flight

Time-of-flight (ToF) technology is based on knowing the time it takes the light to bounce off an object. A light signal is created by a light generating unit and then captured by a sensor as it is reflected by an object. In recent years the ToF has found its use and that led to the creation of ToF cameras. The most popular technology used in ToF cameras is Continuous Wave Intensity Modulation. The scene is constantly illuminated by infrared light, which intensity is modulated so it creates pulses. As an object reflects the light it is then captured by a sensor. This principle can be seen in Fig. 1. Due to the constant speed of light the infrared sensor can determine the distance of a point in space measuring the time it has travelled from the source and back. Device creates a depth map knowing the time shift of the light reflected by an object at different points in space. The resolution of the depth map depends on the resolution of the IR sensor [5].

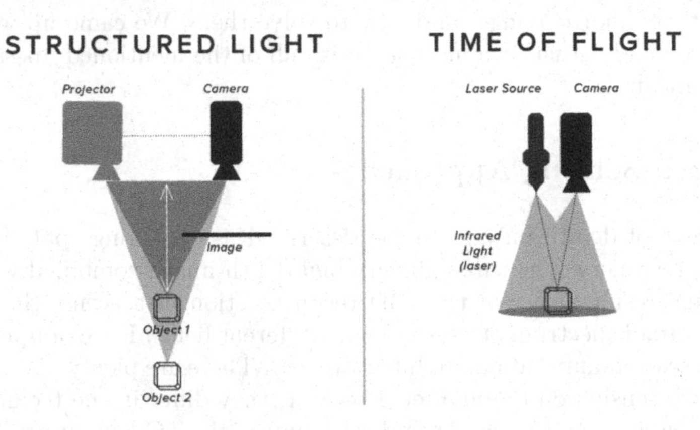

Fig. 1. Principle of structured light and time of flight vision

2.3 Depth Sensing Using Stereoscopic Depth Vision

Active infrared stereoscopic technology works similarly to human vision. Two IR cameras in parallel are used to illuminate an object with IR light. The light has specific textures, which overlap on the object. Both cameras then analyse common points from two different angles. The depth map is then calculated using the spatial shift of these common points. Passive stereo vision combines images from two different angles without the addition of infrared camera. Both technologies can be seen in Fig. 2 [6].

3 Commonly-Available Depth-Sensing Hardware

3.1 Intel RealSense D455

RealSense D445 is a camera with support for depth vision based on technology of stereoscopic vision. Main advantage of this device is its ratio between size, price and its features. Implementation of stereoscopic vision is done by two infrared cameras collecting data of the scene. This data is then sent to a dedicated on-device processor for further processing. Resulting depth image is created by

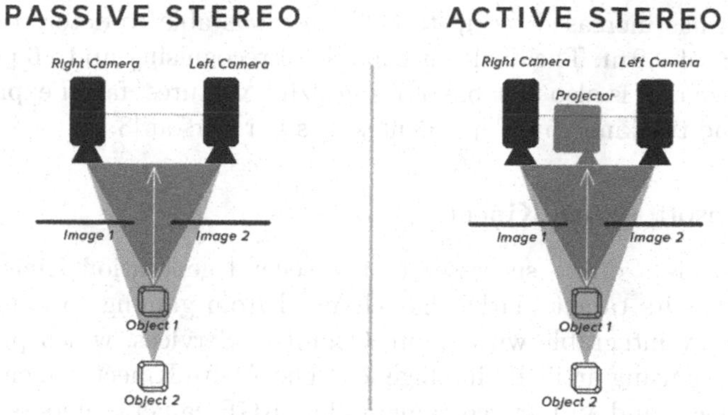

PASSIVE STEREO ACTIVE STEREO

Fig. 2. Principle of active and passive stereo vision

differences of these two images after combining. Resolution of the depth camera is adjustable where maximal resolution is set to 1280 × 720 pixels. Framerate is also adjustable where maximum depth for the camera is set to 90 frames per second. Device is capable of operating in an outdoor environment [7].

3.2 Microsoft Kinect V1

Kinect v1 was the first device from Microsoft which introduced a depth capable device for the user market in 2010. Microsoft's original intention was to use the device in the gaming industry as an add-on to Microsoft Xbox 360 game consoles. The specific implementation is based on two main components, namely an infrared projector operating at a wavelength of 830 nm and a video camera capable of capturing reflection of this information. A specific implementation is based on projecting pseudo-random sequences of points on the scene which are then observed by the infrared sensor. The resulting depth image will be calculated thanks to triangulation. The device is capable of capturing an image with a resolution of 640 × 480 at 30 frames per second. The minimum operating distance reaches 0.5 m and the maximum operating distance with reliable results reaches up to 8 m. User detection is possible at a distance of 1.8 m from the sensor [8,9].

3.3 Microsoft Kinect V2

Kinect v2 is a direct successor to the Kinect v1 available since 2014. The second generation is dramatically different from the first because Microsoft has abandoned structured light technology and introduced ToF technology, which has been detailed above. The device contains an RGB camera, an infrared projector and an infrared receiver. The sensor is capable of capturing depth images with resolution of 512 × 424 and the RGB camera supports a resolution of 1920 × 1080 at a maximum of 30 frames per second. Compared to its predecessor, its depth vision reach has increased by up to 60% and it is able to detect the user at a half distance of 0.9 m. The device is capable of recognising up to 6 people in the sensor's range and is also capable of recognizing gestures, facial expressions and detecting and tracking 25 independent joints per person [8,9].

3.4 Microsoft Azure Kinect

Azure Kinect is a direct successor to the second generation Kinect, with the difference that its target market has changed from gaming to industrial. This device is easily integrable with Azure Cognitive Services, which provide many online services using artificial intelligence. The Azure Kinect camera consists of an RGB camera and an infrared camera. The RGB camera offers several resolution modes where the highest possible resolution is 3840 × 2160 px at 30 frames per second. The infrared camera has the highest resolution of 1024 × 1024 px and uses the ToF principle similarly to its predecessor. In addition, both cameras support different field of view modes. Azure Kinect also has an IMU sensor, consisting of a three-axial accelerometer and a gyroscope, with which the device can estimate its own position in space. Microsoft also offers a Body Tracking SDK that is capable of detecting and tracking the movements of multiple users, each with 32 joints. This SDK is available for Windows and Linux operating systems and the C++ and C# programming languages. Unlike the previous generation of kinetic, current one now support more skeletal joints such as eyes and ears [9,10].

3.5 Comparison of Depth-Sensing Devices

Here we compare the aforementioned camera systems in various aspects. Technical specifications of the systems are summed up in Table 1.

The Influence of Temperature. They found that the Kinect v2 ToF camera because the infrared emitter warms up during operation and needs to be cooled down. The Kinect v1 shows a weak correlation to the temperature. The depth values remain stable and the standard deviation is on an almost constant level. However, in the case of the Kinect v2 the distance measurements exhibit a strong correlation to the temperature. The depth captured varies significantly until the device reaches constant temperature. To achieve stable results using Kinect v2 they recommend to let the device run 25 min before capturing in order to avoid temperature influences.

The Influence of the Camera Distance. In case of Kinect v1 they detected an exponentially increasing offset for increasing distance. While the offset for 0.5 m is below 10 mm, the offset increases more than 40 mm for 1.8 m meaning the image is more precision, but only at close range. Furthermore, it uses a stripe pattern, which is difficult to model and has a huge offset near the edges. On the other hand, Kinect v2 provides an offset of around 18 mm independent of distance. They consider the constant offset to be an advantage in 3D reconstruction applications, since a constant offset can be easily modelled. The central pixels do not deviate, only corner pixels do slightly, due to IR light not illuminating the scene homogeneously.

The Noise. Kinect v2 manages noise per pixel, which implies that less accurate depth measurements appear noisy, because of bigger difference in neighbouring pixels. Furthermore, they mention an artifact called Flying pixel, which is common for ToF cameras. It can occur near edges, image boundaries and depth discontinuities and can also cause noise. On the other hand, Kinect v1 manages noise per-patch appearing less noisy and has no Flying pixels, whilst not using ToF.

Multipath Interference is an effect that occurs when a particular pixel receives light originally sent out for another pixel. It happens commonly when capturing concave surfaces. In the case of Kinect v1 they say the effect is not present. However, for Kinect v2 big offset can be detected in places where the physical properties of the scene cause the light bounce off and interfere with one another. Furthermore, they detected offsets near sharp edges caused by smoothing algorithms.

The Influence of Colour and Surfaces. They say Kinect v1 is not affected by difference in colour. However, Kinect v2 is affected by colour and reflectivity. Darker colours have 10mm higher depth value than lighter colours and have greater deviation. Also, less reflective surfaces are more difficult to capture. Therefore, they suggest using colours with similar reflectivity [11].

Intel RealSense D455 uses stereo cameras, which depth image quality is related to brightness, emitter power, texture, and distance. Brightness of the environment has a negative effect on the image in high contrast scenery. The brighter the environment is, the more power the emitter needs to use. With greater distance or more complex structure the power needs to be also increased. Running at higher power heats up the device and can cause higher noise. Stereo cameras compare two images and calculate depth, therefore the image quality depends on the complexity of the texture. The most prevalent downfall is capturing repetitive textures at further distance, which can cause misinterpretation. Advantage of this camera is it can capture more dense point cloud at 60 fps than Kinect v2 [12,13].

Azure Kinect is similar to Kinect v2 and can be considered an upgraded version. It has a higher resolution, on which it can operate at decent fps. Older generations have rectangular area of capture fitted with valid data. Azure Kinect has a wide field of view, which allows for more compact setup. offers a hex area for narrow field of view and a circular area for a wide field of view. Therefore there are pixels in the rectangular capture that have no information, but the device provides raw, uncropped images. Considering the improved technology and ease to setup Azure Kinect is a best choice for this application [14].

Table 1. Camera specifications.

	RealSense	Kinect v1	Kinect v2	Azure Kinect
Depth sensing technology	Active IR Stereo	Structured light-pattern projection	ToF (Time-of-Flight)	ToF (Time-of-Flight)
RGB camera resolution	1280 × 800 px @ 30 fps	1280 × 720 px @ 12 fps	1920 × 1080 px @ 30 fps	3840 × 2160 px @30 fps
Depth camera resolution	1280 × 720 px @ 90 fps	320 × 240 px @ 30 fps	512 × 424 px @ 30 fps	up to 1024 × 1024 px @ 15 fps
Field of view	87° H, 58° V	57° H, 43° V	70° H, 60° V	up to 120° H 120° V
Measuring distance	0.6–6 m	0.4–4 m	0.5–4.5 m	NFOV unbinned 0.5–3.86 m NFOV binned 0.5–5.46 m WFOV unbinned 0.25–2.21 m WFOV binned 0.25–2.88 m
Number of joints	18	20	25	32

4 Solution

The core principle is to have a room equipped with multiple Azure Kinect devices. A presenter is then placed into the room and captured by those devices. The data collected from all connected Kinect devices will then be further processed. Firstly, a depth representation of the captured scene using the point cloud method is created from each connected device. Subsequently, we detect users in the image using the Body Tracking SDK. These skeletons are then combined into shared space in which we can then attach point cloud representations captured by Kinect devices. The remaining captured points of the scene can be then filtered out as we know where the person is and which points are relevant to him. We can then combine these filtered points into polygons and we can create a vivid dynamic texture or in other words a 3D model in real time for each person. This live 3D model can be then recorded, viewed or streamed over the network.

4.1 Hardware Setup

Our goal is to capture 360 image, and to do that we use Kinect Azure devices. Each Kinect is mounted on a tripod to ensure stability and constant distances of sensors. The configuration is as shown in Fig. 3, two opposing devices facing towards each other with a slight shift to the side. The reason to shift cameras is to avoid interference. We have encountered difficulties setting up the configuration, because the cables supplied in the standard box are not sufficient for this setup.

There is a need to get a 3.5 mm jack cable long enough to connect the two devices. Furthermore, the power cable and the USB cable are both not long enough to connect without extensions. Synchronizing the devices requires a 3.5 jack cable connected to both devices. This way we can set one Kinect to be a master and other one a slave device. Another important step is to set a delay one of the infrared pulse generators in order to prevent misinterpretation of depth data or full on blindness of one sensor due to them being in sync. Then if there is too much noise it is necessary to change the direction of the sensors slightly, until it decreases that noise caused by mutual interference.

HW requirements: Seventh Gen Intel©Core™i5 Processor (Quad Core 2.4 GHz or faster), 4 GB RAM, NVIDIA GeForce GTX 1050 or equivalent, dedicated USB 3.0 port.

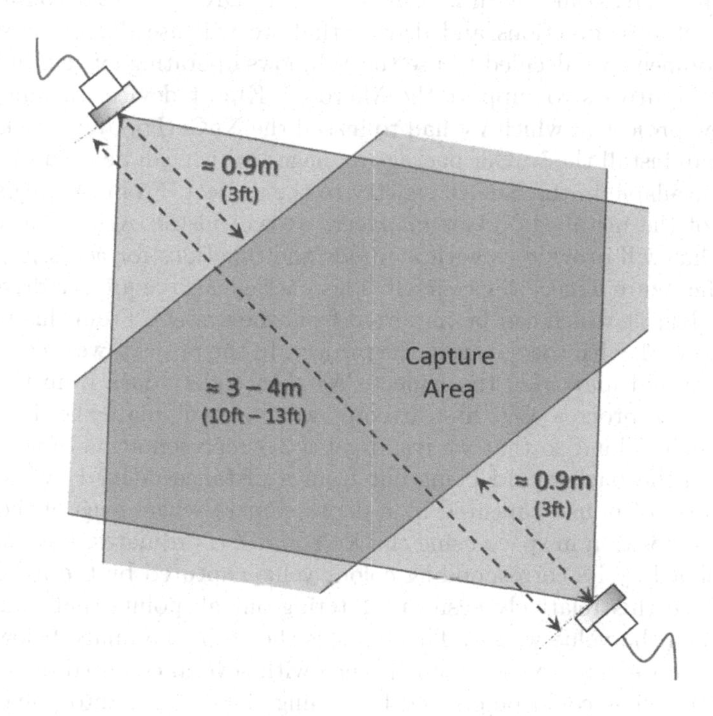

Fig. 3. Example of devices placement

4.2 Unity

Unity is a modern game engine that provides its own development environment. It supports creation of games and applications on many platforms such as PC, mobile devices, game consoles and the web. It also supports game creation in 2D, 3D, AR & VR. Unity currently has several versions available where the basic version is free and the other versions are paid. Free version offers all the basic functionalities for personal use. Paid versions focus on companies and game studios and offer additional services and applications alongside the base version. Unity is also an excellent environment for experimenting and testing new technologies. The basic building block of Unity is a game object that can acquire additional properties such as textures, meshes or scripts. Such an object is then inserted into a scene where it can interact with other objects. This development environment is an ideal tool for our project, as it has native support for three-dimensional space, the C# programming language, and it can also support Azure Kinect DK [15,16].

4.3 Integration with Unity

Azure Kinect DK comes with a multi-platform SDK. The SDK contains a collection of classes, functions and drivers that we will use during development. For development, we decided to use the Windows operating system in which we installed the drivers to support the Microsoft Kinect device. In unity we created a new project in which we had to install the NuGetForUnity package. This package will install the NuGet package manager, which can be used to add other packages available in the NuGet registry to the project. From this registry, with the help of the installed NuGet manager, we will install Azure Kinect Sensor package that will provide us with methods and functions for accessing and controlling the Azure Kinect device itself. This package also requires a driver library for Azure Kinect which can be imported from the system. From this point, it is possible to work with the device in the project. In the project, we created a script for opening and managing the Kinect. We then collect data from this Kinect, which we then process. Our first attempt was to manipulate the data into the so-called point cloud so that we could get a 3D representation of the captured scene, with the background being cut from a distance defined by us. A point cloud is a set of points captured by a depth camera where each of those points has its own position in space using the X, Y and Z coordinates. Each such point is also defined by its corresponding colour value captured by the RGB camera. We managed this relatively easily by filtering out all points that had a higher Z value than the value we set. The result is shown in the image below (Fig. 4). In the image we can see the scanned scene with several coloured dots placed in 3D. A better view could be created by joining these points into polygons, thus creating a dynamic texture in real time.

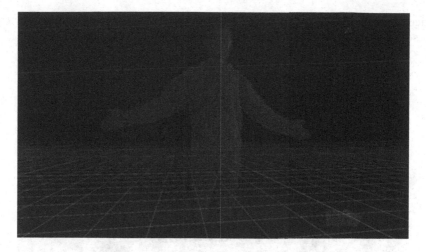

Fig. 4. Example of devices placement

4.4 Body Tracking

Applying a Config Loader script and main body tracking SDK script from Azure Kinect SDK onto a stick figure object we get a visualization of how the person is captured. To map a model of a character we use humanoid animation type in Unity and add the character model into the Unity Scene. We use Kinect4AzureTracker to connect the device to the avatar. Next, we map a stationary part of the human body to a joint on the stick figure. By mapping a pelvis to the root position of Kinect4AzureTracker we get an animated model. Assigning the 3D RGB model to the avatar we create a 3D model of a recognizable person. We can then further modify such a model where, for example, instead of a point cloud representation, we would connect the points to polygons and thus create a dynamic texture in real time [17].

5 Results and Future Work

Utilising Microsoft's innovative approach to their Kinect Azure camera and introducing ready to use SDKs, we were able to digitize a person, and, subsequently, create a 3D model. For one, we made a model using texture created from captured point cloud points in real time. To fill the gaps in the point cloud a mesh of polygons has proven to sufficiently render humanoid representation of a person. Then, we added a colour aspect in sufficient resolution according to the RGB image. Different approach was to attach a pre-existing humanoid model to joints of a body-tracked skeleton. Adding a template RGB image to the model, we created an avatar of a person. Both 3D models can be captured in 30 fps. This data is then serialized and transmitted to server from where it is distributed. Working AR application can then receive this data and display them with ARkit used in devices such as Apple iPhone. ARkit enables our humanoid model to be viewed in a scene through phone camera and observed in AR (Fig. 5).

Fig. 5. Example of the final 3D human model placed in real life environment

Our proposed solution can be used in various industries. In healthcare, for example, a patient can be streamed and observed by a doctor in real time. Another major industry is the gaming industry where people can immerse into a fully virtual world. But one of the main industries we put our efforts in is video conferencing. We plan to continue working on this project in such a way that we will be able to distribute 3D model over the network in real time. The main idea is to create a platform that would allow the connection of multiple users to one conference, in which there would be one presenter. This person would be scanned using Kinect devices, where the complete 3D model would be serialized and compressed on a scanning device from which it would be sent over to the server. The server would then redistribute this model among all connected participants who can join the conference using a generated link. For displaying the resulting model many technologies can be used. Firstly, we would like to address augmented reality using mobile device. Using such a device, the participant could join the conference through the generated link. The participant can then place the received 3D model anywhere in the space.

Acknowledgement. Research described in the paper was financially supported by the 2020-1-CZ01-KA226-VET-094346 DiT4LL ERASMUS+ Innovation Project, MonEd - Modern Trends and New Technologies of Online Education in ICT Study Programs in European Educational Space (KEGA 015STU-4/2021), and the Excellent creative team project VirTel.

References

1. Lapakko, D.: Communication is 93% nonverbal: an urban legend proliferates. Commun. Theater Assoc. Minnesota J. **34**(1), 2 (2007)

2. BIT, T., et al.: A study of learning experience with a dash-based multimedia delivery system. In: EDULEARN 2018: 10th International Conference on Education and New Learning Technologies, EDULEARN Proceedings, Palma, Spain, pp. 8590–8595, 02–04 July 2018. ISSN 2340-1117, ISBN 978-84-09-02709-5, WOS:000531474303006
3. Vančo, M., Minárik, I., Rozinaj, G.: Gesture identification for system navigation in 3D scene. In: 54th ELMAR International Symposium ELMAR-2012, ELMAR Proceedings, Zadar, Croatia, pp. 45–48, 12–14 Sept 2012. ISSN 1334-2630, ISBN 978-953-7044-13-8, WOS:000399723300010
4. Polakovič, A., Vargic, R., Rozinaj, G.: Adaptive multimedia content delivery in 5G networks using DASH and saliency information. In: 25th International Conference on Systems, Signals and Image Processing (IWSSIP), Maribor, Slovenia, 20–22 June 2018. ISSN 2157-8672, ISBN 978-1-5386-6979-2, WOS: 000451277200008, https://doi.org/10.1109/IWSSIP.2018.8439215
5. Sarbolandi, H., Lefloch, D., Kolb, A.: Kinect range sensing: structured-light versus time-of-flight kinect. Comput. Vis. Image Underst. **139**, 1–20 (2015)
6. The basics of stereo depth vision - IntelRealSense™Depth and Tracking Cameras (2021). IntelRealSense™Depth and Tracking Cameras [online]
7. Introducing the IntelRealSense™Depth Camera D455 (2021). IntelRealSense™Depth and Tracking Cameras [online]
8. Zennaro, S., et al.: Performance evaluation of the 1st and 2nd generation Kinect for multimedia applications. In: 2015 IEEE International Conference on Multimedia and Expo (ICME), pp. 1–6. IEEE (2015)
9. Albert, J.A., et al.: Evaluation of the pose tracking performance of the Azure Kinect and Kinect v2 for gait analysis in comparison with a gold standard: a pilot study. Sensors **20**(18), 5104 (2020)
10. Azure Kinect DK depth camera (2021). Docs.microsoft.com [online]
11. Wasenmüller, O., Stricker, D.: Comparison of kinect V1 and V2 depth images in terms of accuracy and precision. In: Chen, C.-S., Lu, J., Ma, K.-K. (eds.) ACCV 2016. LNCS, vol. 10117, pp. 34–45. Springer, Cham (2017). https://doi.org/10.1007/978-3-319-54427-4_3
12. Wen Yu, C.: Stereo-Camera Occupancy Grid Mapping [online]. The Pennsylvania State University (2020). Master thesis. The Pennsylvania State University, Aerospace Engineering. Thesis Advisor: Eric Norman Johnson. [cit. 2021-5-6]. Link: https://etda.libraries.psu.edu/catalog/18031wuc188
13. Gutta, V., et al.: A comparison of depth sensors for 3D object surface reconstruction. In: CMBES Proceedings, vol. 42 (2019)
14. Tölgyessy, M., et al.: Evaluation of the Azure kinect and its comparison to kinect v1 and kinect v2. Sensors **21**(2), 413 (2021)
15. Haas, J.: A history of the unity game engine. Dissertation, WORCESTER POLYTECHNIC INSTITUTE (2014)
16. UNITY TECHNOLOGIES: Unity - Unity (2021). https://unity.com/. Accessed 8 May 2021
17. Shi, S.: Emgu CV Essentials. Packt Publishing Ltd., Birmingham (2013)

Supervised Mixture Analysis and Source Detection from Multimodal Measurements

Johan Lefeuvre[1,2], Saïd Moussaoui[1(✉)], Laurent Grosset[2], Anna Luiza Mendes Siqueira[2], and Franck Delayens[2]

[1] LS2N, Centrale Nantes, CNRS UMR 6004, 1 rue de la Noë, 44321 Nantes, France
said.moussaoui@ec-nantes.fr
[2] CRES, TotalEnergies, Chemin du canal, 69360 Solaize, France

Abstract. This paper presents a method for source detection within unknown chemical mixtures using several spectroscopy measurement modalities. Contrary to the well studied case of single source detection, this approach enables simultaneous detection of multiple chemical components by exploiting the mixing coefficients resulting from supervised linear unmixing and thresholded non-negative least-squares. The first contribution of this work is to propose an automated procedure to compute an optimized binary classifier rule for each component independently using a database of known mixtures. The second contribution is to propose a global decision rule based on the fusion of the multimodal decisions using weighting schemes such as those used in multiple classifier systems (MCS). A real database of Ion Mobiliy Mass Spectrometry (IMMS) data is used to evaluate the detection performance. The main result is to reach an increase of the detection accuracy using the multiple thresholds within the independent classifiers approach as compared to single modality detection.

Keywords: Multimodal supervided spectral unmixing · Sensor fusion · Chemical mixture analysis

1 Introduction

Source detection from spectral data is at the core of several applications of signal processing methods in physical sensing problems, such as chemical substance analysis [6] and hyperspectral imaging [17]. The concept of a source signal is defined as the spectral signature associated to a chemical component.

Classical approaches for source detection and identification are based on the recognition of some discriminant patterns or features of the sought sources [24]. These features can be determined empirically [13, 18] or learned from a large-scale database [10]. Often, the measured signal is interpreted as the linear combination of an unknown set of several sources. In such cases, multivariate analysis techniques [21] such as independent component analysis and non-negative matrix factorization can be used. But, even if these blind separation methods yield

G. Rozinaj and R. Vargic (Eds.): IWSSIP 2021, CCIS 1527, pp. 210–221, 2022.
https://doi.org/10.1007/978-3-030-96878-6_19

estimates of the source and their relative abundances (mixing coefficients), their identification as a physically meaningful components is not guaranteed in all situations.

Alternatively, one can consider a known set of sought sources and estimate the mixing coefficients by a linear regression under non-negativity constraint. These coefficients are then used to make a decision on the presence or the absence of each source in the mixture. In this supervised linear unmixing framework, various methods can be applied, such as thresholded non-negative least-squares [25], non-negative orthogonal greedy algorithms [20] or constrained sparse regression [1]. The main challenge in these approaches is to estimate the sparsity level corresponding to the appropriate number of mixture components. For non-negative least-squares (NNLS), hard thresholding based methods may require a tuning of the threshold level [25] or a setting of an appropriate stopping strategy of the iterative process [2]. For greedy algorithms, the sparsity level should be defined manually or estimated automatically by adopting an adequate stopping rule during the decomposition. In this paper, we explore more specifically the hard thresholding based approach for which we propose two enhancements. The first one is to adopt a source dependent threshold which allows more accurate detection. The second proposal is to use a training database of known mixtures to determine an optimized detection threshold for each sought source separately.

The abundance of each source in an observed mixture data depends on the sensitivity of the sensing modality to this source. Consequently, a lack of sensitivity to some sources in one modality leads to a poor detection of these sources using this modality and conversely an accurate detection will be achieved from another modality more sensitive to these sources. Exploiting multi-modality will therefore enhance detectability of all the sources present in a mixture [7,11]. In the case of multiple measurements of the mixture recorded in the same modality, it has been shown that joint analysis of the data can enhance performance [5,27]. In the case where different measurement modalities are available, an appropriate fusion strategy should be defined. The second contribution of this paper is to adopt a decision fusion method based on a multiple classifier system (MCS) [9]. Finally, the proposed approaches of component-dependent thresholding and multimodal decision fusion are tested on a challenging example of chemical mixture analysis using Ion Mobility Mass Spectrometry (IMMS) data [12,26], where spectral responses of the mixtures and the sources are recorded using two ionization modes and complementary measurement modalities.

2 Problem Statement

Let us consider the case of a single measurement modality. The measurement vector of the mixture is noted $\boldsymbol{y} \in \mathbb{R}^M$, where M represents the number of samples provided by the sensor. Measurement vectors associated to N sources are gathered in a matrix $\boldsymbol{S} = [\boldsymbol{s}_1, ... \boldsymbol{s}_N] \in \mathbb{R}^{M \times N}$. The measurement model is assumed to be a linear mixing:

$$\boldsymbol{y} = \sum_{i=1}^{N} a_i \boldsymbol{s}_i + \boldsymbol{e}, \tag{1}$$

where the additive noise term $\boldsymbol{e}$ corresponds to measurement errors. Computing the vector of mixing coefficients $\boldsymbol{a} = [a_1, ..., a_N]^t \in \mathbb{R}^N$ can be efficiently done by solving the following problem:

$$\hat{\boldsymbol{a}} = \arg\min_{\boldsymbol{a} \in \boldsymbol{C}} \|\boldsymbol{y} - \boldsymbol{S}\boldsymbol{a}\|_2^2, \tag{2}$$

where $\boldsymbol{C}$ denotes the constraint set of the coefficients. For the considered application, measured data correspond to mass spectrometry and ion mobility spectra and are assumed to follow a linear mixing model. The leas- squares problem above (2) is therefore solved under the constraint of non-negativity using a non-negative least-squares algorithm (NNLS) [16] or an interior-point least-squares (IPLS) [4]. The estimated mixing coefficients in each measurement modality are then used to retrieve the detection vector $\boldsymbol{d} = [d_1, ..., d_N]^t \in \{0, 1\}^N$ from $\boldsymbol{a}$ where for $i \in 1, ..., N, d_i = 1$ when component i is present in the mixture and $d_i = 0$ otherwise.

In the case where L independent measurement modalities are available, for each modality $l \in 1, ..., L$, the measurement vector of the mixture is noted $\boldsymbol{y}^{(l)} \in \mathbb{R}^{M_l}$ and the spectra of the N sources are gathered in a matrix $\boldsymbol{S}^{(l)} = [\boldsymbol{s}_1^{(l)}, ...\boldsymbol{s}_N^{(l)}] \in \mathbb{R}^{M_l \times N}$ where M_l is the length of the data vector in the l-th modality. The observation model is then expressed as:

$$\boldsymbol{y}^{(l)} = \sum_{i=1}^{N} \boldsymbol{a}_i^{(l)} \boldsymbol{s}_i^{(l)} \tag{3}$$

with linearly independent abundance vectors $\boldsymbol{a}_i^{(l)}$. The detection of the components which are present in the mixture should therefore be realized by accounting for the values of the mixing coefficients in all modalities using a decision fusion strategy.

Figure 1 shows the proposed detection pipeline. It consists in firstly solving a non-negative linear regression problem in each modality and then deducing the binary detection vectors $(\boldsymbol{d}^{(1)}, ..., \boldsymbol{d}^{(L)})$. The proposed method to compute a fused detection vector $\boldsymbol{d}_{fus}$ from the L unimodal detection vectors $(\boldsymbol{d}^{(1)}, ..., \boldsymbol{d}^{(L)})$ is presented in Sect. 4. Section 5 details an application of the proposed detection method in the case of chemical mixture analysis using IMMS spectrometry data.

3 Detection from Single Modality Measurements

A first step of the proposed approach consists in estimating the mixing coefficients (a_i, for $i \in 1, ..., N$) by solving Problem (2). A NNLS algorithm [16] is used for this purpose. Dedicated optimisation algorithms [4] can be used in order to account for additional constraints (such as sum-to-one). The detection of each mixture components from these mixing coefficient values is addressed as a binary classification problem between two states: presence or absence of each source.

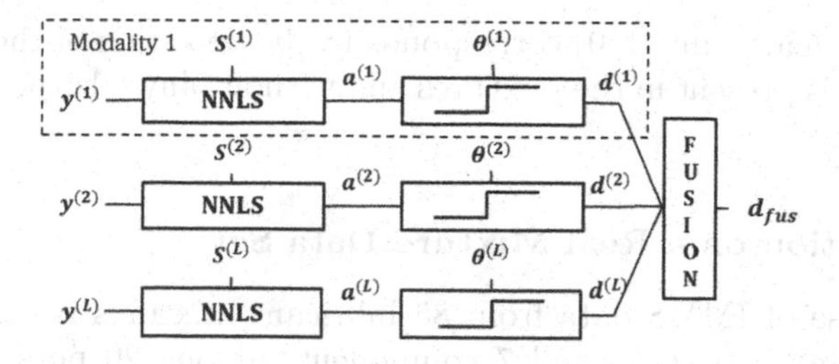

Fig. 1. Multimodal detection pipeline

3.1 Detection Approach

The decision rule related to the two states leads to a binary classifier $h_{\theta_i}(a_i) : \mathbb{R} \to \{0, 1\}$ according to:

$$h_i(a_i; \theta_i) = \begin{cases} 1 \text{ if } & a_i \geq \theta_i \\ 0 & \text{otherwise} \end{cases} \tag{4}$$

Our proposal consists in adopting thresholds θ_i which are specific to each source in the considered measurement modality. It allows to account for the different sensitivities of the measurement modalities on the sought sources. The detection vector is then given by $d_i = h_i(a_i; \theta_i)$ for $i \in 1, ..., N$. The values of the thresholds θ_i are defined by optimization of the detection performance on a training database of known mixtures. The best detection threshold values are chosen in such a way to reach a balance between sensitivity and specificity of the binary classifier.

3.2 Specification of the Detection Thresholds

Let $\boldsymbol{Y} = [\boldsymbol{y}_1, ..., \boldsymbol{y}_K] \in \mathbb{R}^{M \times K}$ denote a matrix of K mixtures with a subset of known sources among a set of N sources. The detection performance for each binary classifier corresponding to the i-th source are evaluated in terms of true positive rate and false positive rate

$$\text{TPR}_i\ (\theta_i) = \frac{\text{TP}_i}{\text{TP}_i + \text{FN}_i} \quad \text{and} \quad \text{FPR}_i(\theta_i) = \frac{\text{FP}_i}{\text{FP}_i + \text{TN}_i}, \quad \text{for } i = 1, \ldots, N,$$

where TP (True Positive) and FP (False Positive), correspond to the number of times that sources are detected as present by the binary classifier and they are actually present (resp. absent) in the mixture. TN (False Negative) and FN (False Negative) correspond to the cases where components are detected as absent by the binary classifier and they are actually absent (resp. present) in the mixture. The best compromise between sensitivity (TPR) and specificity (1-FPR) is achieved by defining the threshold values according to

$$\hat{\theta}_i = \arg\min_{\theta_i \in \mathbb{R}} \|(\text{TPR}_i(\theta_i), \text{FPR}_i(\theta_i)) - (1, 0)\|_2^2. \tag{5}$$

Note that the ideal point $(1, 0)$ corresponds to the detection of the source (sensitivity) when it is present in the mixtures and without any false positive detection (specificity).

3.3 Illustration on a Real Mixture Data Set

A real database of IMMS data from 85 lubricant mixtures is considered. Each mixture contains between 5 and 7 components among 20 possible ones. More details on this data set are given in Sect. 5. Figure 2-(a) gives the distance between the receiver operating characteristic (ROC) curve for the detection of different sources. It can be noted that the optimal detection thresholds for the considered sources (C6, C15 and C17) are different, which suggests to use a source dependent threshold. Moreover, the global threshold seems to be appropriate for C15 but it is not optimal for the two other sources (C6 and C17).

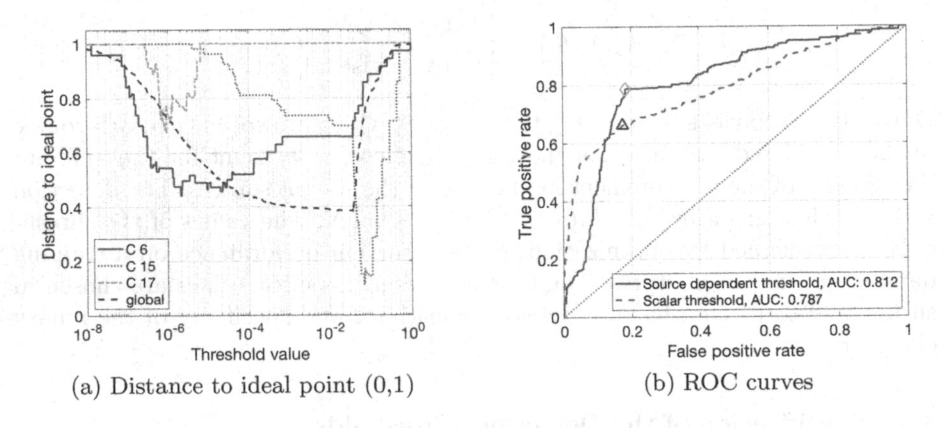

(a) Distance to ideal point (0,1) (b) ROC curves

Fig. 2. Influence of the threshold values on the detection performance

Figure 2-(b) shows a comparison between the average ROC curve obtained when applying a source dependent threshold and the ROC curve obtained by using the same threshold for all the sources. Both the Area Under Curve (AUC) and the average performance at the optimal point are higher in the case of source dependent thresholds.

4 Detection Strategy in the Multimodal Case

This section addresses the detection in the case where $L > 1$ independent measurement modalities are available. To illustrate the relevance of this strategy let us compare the ROC curves, shown in Fig. 3, obtained with two different sources C1 and C3 in the four modalities offered by the IMMS spectrometer. The ROC curves of modality M2 (to the left) yields the best performance and M3 the worst performance for source C1. In contrast, for component C3, the best detection is obtained by modality M1 and the worst detection is obtained with modality M4. This example illustrates the complementarity of the different modalities and the need for a decision fusion strategy for an accurate detection of the entire set of sources.

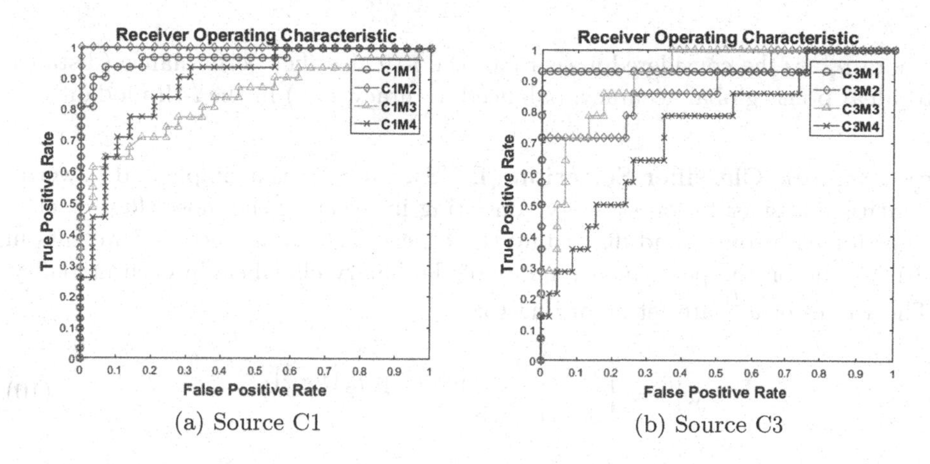

(a) Source C1 (b) Source C3

Fig. 3. ROC curves for source detection in four modalities

4.1 Weighting Schemes

The fusion of the independent binary classifiers $(h_i^{(1)}, \ldots, h_i^{(L)})$ in the L modalities is performed using a linear combination rule according to

$$g_i\left(a_i^{(1)}, \ldots a_i^{(L)}\right) = \sum_{l=1}^{L} \omega_i^{(l)}\, h_i^{(l)}\left(a_i^{(l)}; \theta_i^{(l)}\right), \tag{6}$$

where $\theta_i^{(l)}$ and $\omega_i^{(l)}$ correspond to the detection threshold and the decision weight associated to the detection of i-th component in the l-th measurement modality. A resulting fused decision corresponds to a Multiple Classifier System [9] defined subsequently as:

$$d_i^{\text{fus}} = \begin{cases} 1 & \text{if } g_i\left(a_i^{(1)}, \ldots a_i^{(L)}\right) \geq 0.5 \\ 0 & \text{otherwise} \end{cases} \tag{7}$$

Depending on the values assigned to the decision weights, one can distinguish mainly three different fused classifiers.

a) Majority Vote (MV). It consists of retaining the decision taken by the absolute majority of classifiers [3]. It is defined as below:

$$\omega_i^{(l)} = \frac{1}{L}. \tag{8}$$

However, the MV classifier doesn't account for the performance of the binary classifiers in the different measurement modalities.

b) Weighted Majority Vote (WMV). Weighted Majority Vote [15], [19] is a decision fusion rule in the case of independent classifiers and the weights are defined as:

$$\omega_i^{(l)} = \log \frac{\mu_i^{(l)}}{1 - \mu_i^{(l)}} \tag{9}$$

where $\mu_i^{(l)}$ is the considered measure of the i-th classifier performance (distance to ideal point, global accuracy, balanced accuracy, etc.) in the l-th modality.

c) Dynamic Classifier Selection (DCS). DCS [8] is a simple and powerful multiple classifier fusion strategy consisting in selecting the most efficient classifier for each source and discarding the others. The chosen decision weights in (6) depend on the performance index of the binary classifiers in each modality. The values of $\omega_i^{(l)}$ are set according to:

$$\omega_i^{(l)} = \begin{cases} 1 & \text{if } \mu_i^{(l)} = \max_{k \in [\![1,L]\!]}\{\mu_i^{(k)}\}. \\ 0 & \text{otherwise.} \end{cases} \tag{10}$$

4.2 Performance Index

The weighting schemes for decision fusion are based on classifier performance. The most commonly used performance measure is the detection accuracy (ACC), defined as:

$$\mu_i^{\text{acc}} = \frac{\text{TP}_i + \text{TN}_i}{\text{TP}_i + \text{TN}_i + \text{FP}_i + \text{FN}_i}. \tag{11}$$

This index is commonly used for evaluation of classifiers with balanced occurrences of presence/absence of the sources.

In the case of our application, the source component are more often absent than present in the mixtures which leads trivial classifiers with high rejection rate to yield good global accuracy scores. This phenomenon called "curse of accuracy" [14] is avoided by choosing a more adapted performance index such as the Mathews Correlation Coefficient (MCC) [23], which is defined as

$$\mu_i^{\text{mcc}} = \frac{\text{TP}_i \times \text{TN}_i - \text{FP}_i \times \text{FN}_i}{\sqrt{(\text{TP}_i + \text{FP}_i)(\text{TP}_i + \text{FN}_i)(\text{TN}_i + \text{FP}_i)(\text{TN}_i + \text{FN}_i)}}. \tag{12}$$

This index is considered as one of the best binary classifier performance metrics since it is not affected by class imbalances in the training set. An ideal classifier leads to $\mu_i^{\text{mcc}} = 1$, a random classifier gives $\mu_i^{\text{mcc}} = 0$ whether or not the training set is balanced, meanwhile a classifier systematically predicting the exact opposite of the ground truth will hace an MCC value $\mu_i^{\text{mcc}} = -1$.

5 Application to Mixture Analysis by IMMS Spectrometry

In this section, detection performance of the presented MCS are compared to the performance obtained separately on each modality on a supervised chemical mixture analysis by an IMMS spectrometer.

5.1 Mixture Synthesis

A database of 85 mixture made from 20 different chemical components has been designed. Each mixture contains 5, 6 or 7 components randomly chosen from a set of 20 classical components involved in lubricant formulation [22]. Samples from each mixture and each source components have been analysed twice with an IMMS spectrometer [12]. This spectrometer ionizes the analyte sample to create a swarn of ions. In a second step, bidimensional maps corresponding to distribution of drift times and time of flight of the ions through two separation chambers are recorded. The distribution of drift time through the first chamber is called the ion mobility spectrum and the distribution of time of flight in the second chamber leads to the mass spectrum. Mass spectra and ion mobility spectra are considered as the spectral signatures of the analyte sample.

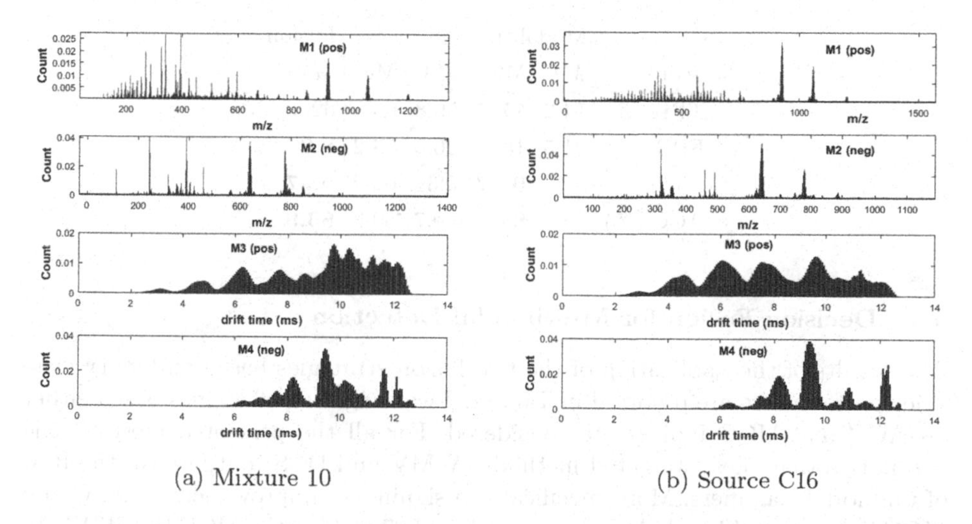

(a) Mixture 10 (b) Source C16

Fig. 4. Spectral responses of a mixture and one source in four measurement modalities

Two distinct ionization modes have been used, therefore 4 one-dimensional spectra are recorded for each sample. Those modalities are called M1, M2, M3 and M4, corresponding to positive ionisation mass spectra, negative ionisation mass spectra, positive ion mobility spectra and negative ion mobility spectra. The spectral signatures of one of the mixture are presented in Fig. 4-(a). Figure 4-(b) shows the spectral signatures of one component of this mixture.

5.2 Mixture Analysis

The IPLS algorithm [4] is used for the mixing coefficient estimation under non-negativity constraint. Among the 85 observed mixtures, a set of randomly chosen 60 mixtures, with balanced occurrences of each component, are retained for training the algorithms in terms of optimal threshold values setting and decision weights calculation. The remaining set of 25 observed mixtures are retained for the performance evaluation. This procedure is repeated with 60 independent realizations to get statistically robust global performance indexes.

5.3 Unimodal Detection and Majoruty Vote

Global performance of each optimized unimodal classifier and the majority vote of the 4 modalities on the database are presented in Table 1. It can be noted that modality M2 seems to present the best global performance in terms of true positive rate, false negative rate and overall accuracy. It can be noted that a naive global fusion scheme such as MV leads to an improvement of specificity (lower TPR) but at the expense of sensitivity (higher FPR). This results suggests adopting alternative fusion strategies

Table 1. Performance of unimodal detection

	Modalities				Fusion
Score	*M1*	M2	M3	M4	*MV*
TPR (%)	74.2	**77.7**	59.8	63.8	62.5
FPR (%)	21.5	18.4	26.3	29.2	**8.2**
MCC	0.50	**0.57**	0.31	0.32	**0.57**
ACC (%)	77.4	80.8	69.7	69.3	**83.9**

5.4 Decision Fusion for Multimodal Detection

The results of the application of the two fusion strategies based on binary classifier performance are reported in Table 2. Two weighting strategies based either on ACC and MCC indices are considered. For all the presented metrics, one can notice that these weighted methods (WMV and DCS) outperform the best of unimodal classifiers. More specifically, a significant improvement in ACC and MCC indices is achieved. An accuracy value of 87.9 % for the MCC based WMV. It notably reduced the rate of false positives (5.1%) while being less sensitive than the modality M2. However, the MCC based DCS classifier is the most sensitive one (82.3 %) and presents a moderate rate of false alarms (13.2 %).

5.5 Discussion on Component Detection

Figure 5 shows the detection results of two groups of sources. In the first group, the sources are well detected by positive ionization mode while in the second group, the sources are better detected using negative ionization mode. For each group of components, unimodal detection results are compared to the multi-modal detection. In both cases, the performance of the fused decision is equivalent to the best of all the unimodal classifiers. The proposed framework has therefore been able to fully exploit the complementarity of the modalities.

Table 2. Performance of fusion strategies based on either ACC or MCC measures.

	ACC		MCC	
Score	WMV	DCS	WMV	DCS
TPR (%)	77.3	81.5	68.3	**82.3**
FPR (%)	11.8	12.5	**5.1**	13.2
MCC	0.65	**0.67**	0.62	**0.67**
ACC (%)	85.5	86.0	**87.9**	85.6

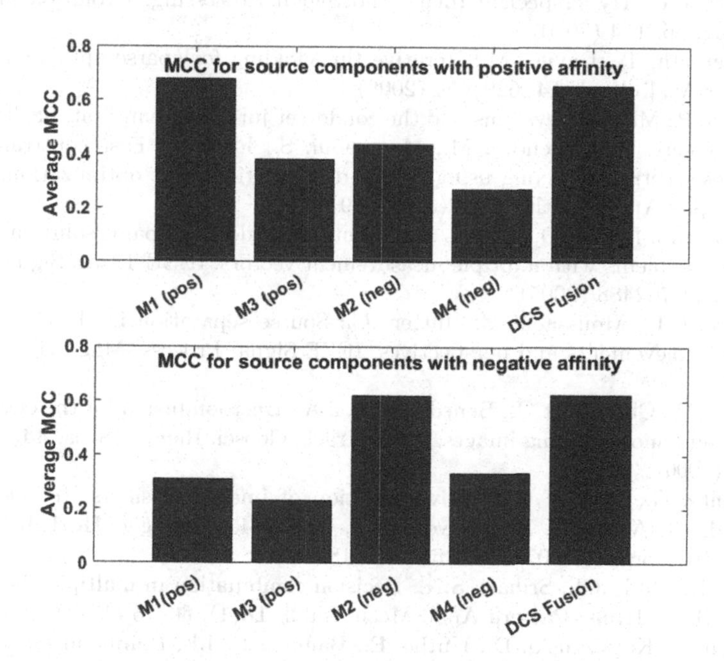

Fig. 5. Detection performance using measurements with two ionization modes

6 Conclusion

The concrete problem of identifying chemical components in an unknown mixture from multi-model spectrometry data, while being an instance of the well-known source separation problem, poses many challenges among which the source-dependent response sensitivity and need to derive robust decision fusion strategies to combine information provided by the different modalities. It has been shown on real data that a more accurate detection is achieved by proposing a component-specific threshold in each modality and adopting a decision fusion scheme exploiting the detection performance in each modality, measured in a training database. We also noted that DCS classifiers tend to be very sensitive while WMV are more specific. Future works will be directed at investigating methods based on greedy sparse recovery, proposing adequate rules for the joint decomposition of the observed data with multiple modalities. Another perspective proposal of decision rules based on machine learning approaches and that will not requite the linear mixing model hypothesis.

References

1. Bioucas-Dias, J.M., Figueiredo, M.A.T.: Alternating direction algorithms for constrained sparse regression: application to hyperspectral unmixing. In: 2010 2nd Workshop on Hyperspectral Image and Signal Processing: Evolution in Remote Sensing, pp. 1–4 (2010)
2. Blumensath, T., Davies, M.: Iterative thresholding for sparse approximations. J. Fourier Anal. Appl. **14**, 629–654 (2008)
3. Boland, P.: Majority systems and the condorcet jury theorem. Stat. **38**, 181 (1989)
4. Chouzenoux, E., Legendre, M., Moussaoui, S., Idier, J.: Fast constrained least squares spectral unmixing using primal-dual interior-point optimization. IEEE J. Sel. Topics Appl. Earth Obs. **7**(1), 59–69 (2014)
5. Cotter, S.F., Rao, B.D., Engan, K., Kreutz-Delgado, K.: Sparse solutions to linear inverse problems with multiple measurement vectors. IEEE Trans. Signal Process. **53**(7), 2477–2488 (2005)
6. Duarte, L.T., Moussaoui, S., Jutten, C.: Source separation in chemical analysis: recent achievements and perspectives. IEEE Signal Process. Mag. **31**(3), 135–146 (2014)
7. Fauvel, M., Chanussot, J., Benediktsson, J.A.: Decision fusion for the classification of urban remote sensing images. IEEE Trans. Geosci. Remote Sens. **44**(10), 2828–2838 (2006)
8. Giacinto, G., Roli, F.: Adaptive selection of image classifiers. In: Del Bimbo, A. (ed.) ICIAP 1997. LNCS, vol. 1310, pp. 38–45. Springer, Heidelberg (1997). https://doi.org/10.1007/3-540-63507-6_182
9. Ho, T.K., Hull, J.J., Srihari, S.N.: Decision combination in multiple classifier systems. IEEE Trans. Pattern Anal. Mach. Intell. **16**(1), 66–75 (1994)
10. Horsch, S., Kopczynski, D., Kuthe, E., Baumbach, J.I., Rahmann, S., Rahnenfü, J.: A detailed comparison of analysis processes for MCC-IMS data in disease classification-automated methods can replace manual peak annotations. PLOS ONE **12**(9), 1–16 (2017)
11. Jeon, B., Landgrebe, D.A.: Decision fusion approach for multitemporal classification. IEEE Trans. Geosci. Remote Sens. **37**(3), 1227–1233 (1999)
12. Kanu, A.B., Dwivedi, P., Tam, M., Matz, L., Hill, H.H., Jr.: Ion mobility-mass spectrometry. J. Mass Spectrom. **43**(1), 1–22 (2008)
13. Kopczynski, D., Baumbach, J.I., Rahmann, S.: Peak modeling for ion mobility spectrometry measurements. In: Proceedings of the 20th European Signal Processing Conference (EUSIPCO), pp. 1801–1805 (2012)

14. Kubat, M., Matwin, S., et al.: Addressing the curse of imbalanced training sets: one-sided selection. In: ICML, Nashville, USA, vol. 97, pp. 179–186 (1997)
15. Kuncheva, L.: Combining Pattern Classifiers: Methods and Algorithms. John Wiley & Sons, Hoboken (2014)
16. Lawson, C.L., Hanson, R.J.: Solving Least Squares Problems. SIAM, Philadelphia (1995)
17. Manolakis, D., Shaw, G.: Detection algorithms for hyperspectral imaging applications. IEEE Signal Process. Mag. **19**, 29–43 (2002)
18. Marczyk, M., Polanska, J., Polanski, A.: Improving peak detection by gaussian mixture modeling of mass spectral signal. In: 2017 3rd International Conference on Frontiers of Signal Processing (ICFSP), pp. 39–43 (2017)
19. Moreno-Seco, F., Iñesta, J.M., de León, P.J.P., Micó, L.: Comparison of classifier fusion methods for classification in pattern recognition tasks. In: Yeung, D.-Y., Kwok, J.T., Fred, A., Roli, F., de Ridder, D. (eds.) SSPR /SPR 2006. LNCS, vol. 4109, pp. 705–713. Springer, Heidelberg (2006). https://doi.org/10.1007/11815921_77
20. Nguyen, T.T., Idier, J., Soussen, C., Djermoune, E.: Non-negative orthogonal greedy algorithms. IEEE Trans. Signal Process. **67**(21), 5643–5658 (2019)
21. Pomareda, V., Calvo, D., Pardo, A., Marco, S.: Hard modeling multivariate curve resolution using lasso: application to ion mobility spectra. Chemom. Intell. Lab. Syst. **104**(2), 318–332 (2010)
22. Ponthus, J., Riches, E.: Evaluating the multiple benefits offered by ion mobility-mass spectrometry in oil and petroleum analysis. Int. J. Ion Mobility Spectrom. **16**(2), 95–103 (2013)
23. Powers, D.: Evaluation: from precision, recall and F-factor to ROC, informedness, markedness & correlation. J. Mach. Learn. Technol. **2**, 37–63 (2011)
24. Scharf, L.L., Friedlander, B.: Matched subspace detectors. IEEE Trans. Signal Process. **42**(8), 2146–2157 (1994)
25. Slawski, M., Hein, M.: Sparse recovery by thresholded non-negative least squares. In: Proceedings of the 24th International Conference on Neural Information Processing Systems (NIPS 2011), pp. 1926–1934. Curran Associates Inc., Red Hook (2011)
26. Szymańska, E., Davies, A.N., Buydens, L.M.: Chemometrics for ion mobility spectrometry data: recent advances and future prospects. Analyst **141**(20), 5689–5708 (2016)
27. Tropp, J., Gilbert, A., Strauss, M.: Algorithms for simultaneous sparse approximation. Part i: greedy pursuit. Signal Process. **86**, 572–588 (2006)

Author Index

Printed in the United States
by Baker & Taylor Publisher Services

Printed in the United States
by Baker & Taylor Publisher Services